Radio Society of Great Britain

Centenary

100 Years Working for Amateur Radio

by Elaine Richards, G4LFM

Published by the Radio Society of Great Britain, 3 Abbey Court,
Priory Business Park, Bedford MK44 3WH, United Kingdom

Published 2013

© Radio Society of Great Britain, 2013. All rights reserved. No part of this publication may be reproduced, stored in a retrieval system, or transmitted, in any form or by any means, electronic, mechanical, photocopying, recording or otherwise, without the prior written permission of the Radio Society of Great Britain.

ISBN: 9781 9050 8689 4

Publisher's note
The opinions expressed in this book are those of the author and not necessarily those of the RSGB. While the information presented is believed to be correct, the author, the publisher and their agents cannot accept responsibility for consequences arising from any inaccuracies or omissions.

Cover design: Kevin Williams
Typography and design: Mark Pressland
Production: Mark Allgar, M1MPA

Printed in Great Britain by Berforts Information Press

Contents

Foreword v

1 In the Beginning 1

2 Early Experiments, Famous Firsts & Pioneers 7

3 Inter War Years 23

4 World War II 55

5 Post War to Century's End 67

6 New Millennium 117

   Appendix 137

   Index 138

Thanks to Martin Atherton G3ZAY for his help in cataloguing and scanning the archives and to Darryl Simpson M6WAS and Ken Hatton G3VBA for their transcribing of handwritten letters.
Thanks most of all to the family who have helped by reading the text, kept the coffee hot and understood how much I'm enjoying this research.

Elaine Richards, G4LFM

Introduction

In its first 100 years, the Radio Society of Great Britain (RSGB) has represented the interests of Members at national and international level. It has encouraged technological breakthroughs, propagation experiments and scientific exploration. Whether that has involved proving the short wave bands could be used for long distance communication, bouncing signals off the moon or evaluating communication during a solar eclipse, the RSGB and its Members have been involved in it all. The Society provided the framework within which the early pioneers and experimenters thrived. One hundred years later, it still provides a platform for the exchange of ideas amongst experimenters and, although the technologies may have changed, the amateur spirit has not.

Amateur radio has changed dramatically in this first 100 years and so has the Society. From a Membership of four to the tens of thousands of today, the RSGB has changed and moved with the times creating a modern Society ready for the next 100 years.

The history of the RSGB is a rich tapestry indeed. The stories gathered together here provide a glimpse of the people behind the organisation and the way they have shaped the history of the Society. Not all the stories, famous names and events have made it into print. Nevertheless, it is hoped that this collection of events will give a flavour of RSGB history and the hobby it supports.

Elaine Richards, G4LFM
May 2013

Centenary

In The Beginning

Three of the founder members, Rene Klein (seated), L F Fogarty (left), Leslie McMichael (centre) and Frank Hope-Jones (right) who became the first Chairman.

LETTER TO THE EDITOR

The London Wireless Club started after Rene Klein had a letter published in *English Mechanic* on 6 June 1913. He expressed surprise that there wasn't a wireless group in London when other major cities had such gatherings. A meeting of interested parties took place on 5 July and the London Wireless Club – with a membership of four – was born. The founding members were Rene Klein, Leslie McMichael, L Francis Fogarty and A P Morgan. The annual subscription was set at 10s 6d for town members and 5s for country members, showing that there was an intention from the very beginning to expand beyond London.

Following that first meeting, a notice appeared in both *English Mechanic* and *Wireless World* inviting interested parties to get in touch and join Klein for a general meeting in September when a proper committee would be elected. There was a good response to these notices and it was at this general meeting that Frank Hope-Jones joined the Society. He suggested that the Society should work with the authorities to negotiate a "charter of freedom for the wireless enthusiast". He also suggested "the appointment of men of eminence in the science of wireless telegraphy as Vice Presidents, from whom a President should be selected". Sadly, several of the first Vice Presidents died during World War I (WWI) and when the Society reconvened after the war, several other

Centenary

Rene Klein G8NK,
Founder of the Radio Society of Great Britain on 5 July 1913.

Leslie McMichael G2MI
and his Membership card No. 1.

1. In The Beginning

The Founders

Rene Klein G8NK

Rene Klein, (licensed as RKX in 1913) was the first Honorary Secretary of the Society. Klein was licensed as 2HT and later as G8NK by which he was well known on the bands. He operated regularly from the room in which the Society was formed. During October 1913, the GPO authorised Klein to transmit signals from his home to a station set up at the Model Engineer Exhibition being held in the Horticultural Hall, Westminster from the 12th to 18th of that month. Transmissions were made on a wavelength of 250m and with an input of 40W. This is the first record of an exhibition station. In 1915, he wrote to The Times suggesting that members of the Society could keep watch for illicit wireless transmission. Although the idea wasn't taken up at the time, during WWII many radio amateurs became Voluntary Interceptors. Klein was elected Vice President shortly after WWI and made an Honorary Member in February 1954.

Leslie McMichael G2MI

Leslie McMichael (licensed as MXA in 1913, 2MI in 1919 and later as G2FG) began experimenting with wireless in 1903 using a 10 inch spark coil. He sent signals about 200 yards and received them on a home made coherer made from nickel and silver filings in a glass tube. During WWI, he served in the Wireless Instructional Section of Royal Flying Corps. In July 1919, McMichael took over the role of Secretary of the Society, doing much of the work that culminated in the re-issue of licences. Then in 1921 he founded the radio firm McMichael Radio. He was Secretary of the Society until 1924. In 1945 he was made an Honorary Member.

L Francis Fogarty FFX

L Francis Fogarty (licensed as FFX) became the Society's first Treasurer, a post he retained until 1924 when the pressures of business with his newly founded Zenith Electrical Company became too great. He was part of the deputation that went to the Postmaster General in 1921 to ask for half-hour public broadcasts to be permitted from Writtle, near Chelmsford; these were the forerunners to broadcasts from the BBC.

A P Morgan

A P Morgan demonstrated one of the smallest working loud speaker 3-valve receivers made from purchasable components at one of the very early meetings. Little else is known of this founding Member.

Frank Hope-Jones

Frank Hope-Jones, the Society's first Chairman, was the person who initiated the move that led many other societies to seek affiliation to the Wireless Society of London. He described how, if clubs affiliated, it would enable the Society to represent, effectively, the views of a large body of amateurs in licensing matters. He was a non-transmitting licence holder.

A group of officers of the Society, taken in October 1919. Back row left to right shows L F Fogarty, R H Klein, L McMichael, E W Kitchen, M Child. Front row left to right, Dr Erskine-Murray, A A Campbell Swinton, Admiral Sir Henry Jackson, F Hope-Jones.

eminent names were offered Vice Presidency. Names such as Sir Charles Bright, Prof Ernest Fleming, General Ferrié and Mr S G Brown joined the Society and Sir Oliver Lodge and Marconi accepted Honorary Membership.

W Lloyd G5TV, who became a Colonel in the Royal Signals, could have been one of the founder members. He had written to Rene Klein following the letter in the *English Mechanic* and was invited along to the first meeting. At the time he wasn't even 17 years old – his birthday was a month after that first meeting. On arriving at Rene Klein's house he was rather overawed by the "bearded technical gentlemen all old enough to be my father" attending the meeting, so he left without going inside and returned home. His own father told him off for his lack of courage! Lloyd did write to Rene Klein again and received an invitation to join the London Wireless Club – the letter arrived on his 17th birthday.

Just ten weeks later the membership had increased to 40. At the General Meeting called on 23 September 1913, the name was changed to the Wireless Society of London. Two grades of membership were created – Members and Associate Members. Full membership was restricted to persons over the age of 21 who had been engaged in experimental work for at least two years and / or who had satisfied the committee that they possessed the necessary qualifications or training. By the end of the first year there were 151 Members and 11 Associate Members – 69 of whom held transmitting licences.

Shortly after the formation of the Society, an event occurred that showed the need for an organisation with much wider horizons and aims. The General Post Office (GPO) announced that, for the first time, it intended to introduce a charge (one guinea) for issuing transmitting and receiving permits, although it was stated that this would be an initial fee only and that there would be no annual charge. Amateurs felt this contravened the spirit of the 1904 Act that specifically safeguarded the rights of 'experimenters'. Klein asked for a meeting to discuss the new regulations. So, in July 1913, a meeting took place that began the history of official liaison between the Society and the licensing authorities – a practice that continues to this day.

By the end of 1914, the Society had 212 Members, 31 Associate Members and two Foreign Members on the books.

WORLD WAR I

On 1 August 1914, at the outbreak of WWI, the GPO took possession of 'all the stations established by the Licensees for the working of telegraphy'. This was to ensure that 'His Majesty's Government shall have control over the transmission of messages'. Equipment was taken into Post Office custody and was only returned after the end of hostilities when the amateurs were expected to re-apply for their licences.

Amateurs were told, as a measure of safety, to 'remove at once your aerial wires and dismantle your apparatus'. So started more than six years of a lack of amateur radio experimentation in the UK.

The Society could be justly proud of its war services. Its younger Members, ready trained, flocked to the wireless units of the Navy, Army and Air Forces, where they served with distinction. Those who stayed at home set a good example by shutting down their own installations.

During the war, Society Members filled many important posts in the wireless sections of all three services. For example, Simmonds 2OD joined the Wireless Section of the Royal Engineers where he was "afforded considerable opportunity for experimental work with thermionic valves of the types then available, both transmitting and receiving...and this experience formed a ground work for the experimental work undertaken after the war."

War did nothing to slow the development of wireless technology. There was a great deal of effort invested in improving communications equipment. Probably the greatest step forward during this time was the development of the superheterodyne receiver. This was developed by Edwin Armstrong, an American electrical engineer and inventor. He patented the regenerative circuit in 1914, the superheterodyne receiver in 1918 and the super-regenerative circuit in 1922.

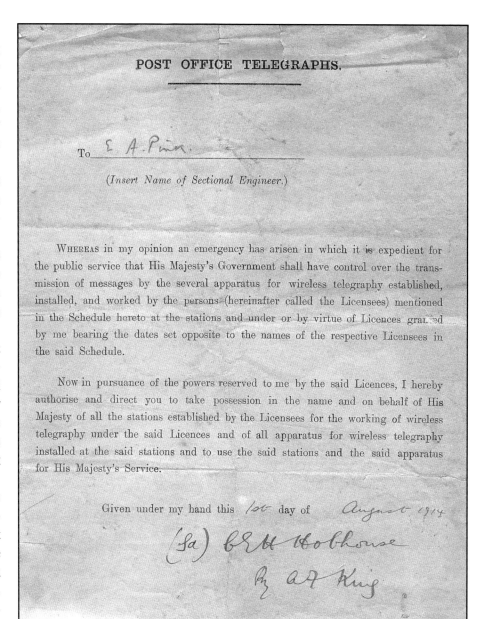

Instruction to the Sectional Engineer of Post Office Telegraphs, E A Pink, to take possession of wireless telegraphy equipment, dated 1 August 1914.

H W Pope's pre-war station that was dismantled by the GPO. Pope's callsign was PZX.

BACK ON THE AIR

Receiving licences were granted in October 1919 and amateur licenses could be applied for in November 1919, although it had needed support by the likes of Marconi, Eccles and Fleming before this came about. A receiving licence was reasonably easy to obtain. You needed to supply proof of British nationality, two written references and an undertaking to observe the secrecy of messages.

Gradually the restrictions on components were relaxed and electrical buzzers, spark coils and headphones could be purchased. Spark coils and headphones came with a signed declaration that they would not be used for the sending or receiving of messages by wireless telegraphy, except with the written permission of the Postmaster General. Thermionic valves couldn't be used without special authority.

The new transmitting licences were surrounded with several restrictions including "communication will be authorised only with specified stations and not exceeding five in number". Power was limited to 10 watts except with special permission and it was just on the 1000 and 180 metre bands with other restrictions around the hours of working. Artificial aerial licences were issued to experimenters not requiring the use of a radiation licence. You needed to show the scientific value or general public benefit your transmissions would achieve to be able to transmit. The new callsigns were, at first, the number 2 followed by two letters, although the numbers 5 and 6 were also used. Many of the new calls were based on initials or names.

The Society continued to work with the GPO to gradually reduce the restrictions, but there were few major concessions until after World War II.

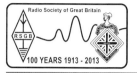

Early Experiments, Famous Firsts & Pioneers

BROADCASTING ORGANISATION

Once licences were restored, broadcasting soon got under way, particularly in the London area where few evenings in the week were without entertainment of some kind. So much so that the Society called the holders of transmitting licences together in July 1921 to regulate their programmes in order to avoid interference. All this was on low power and, technically, because it was broadcasting entertainment, it was against the law. In March 1921, the President, Dr J Erskine Murray, asked that permission should be granted to the Marconi Company to transmit a weekly concert. It took six months for the authorities to make up their mind – the answer was no. On 29 December 1921 the Postmaster General received a petition from the Society signed by the Presidents of 63 Societies around the UK requesting permission to broadcast. Within a fortnight, permission was granted for the station 2MT based at Writtle in Essex to begin its Tuesday evening concerts. This eventually led to a National Broadcasting service – the BBC.

Dr Erskine-Murray.
RSGB President 1921

RSGB EMERGES

In November 1922, the Wireless Society of London changed its name for the last time to the Radio Society of Great Britain. It described itself thus, "The Radio Society of Great Britain is a virile and progressive body of amateur radio experimenters bonded together for the promotion of knowledge and brotherhood of those interested in radio art. It exists also with the object of the advancement of the art, the representation of the amateur in legislative matters and for the disciplined use of the ether in so far as amateur experimenters are concerned".

LONG DISTANCE COMMUNICATIONS

Great advances were being made in radio communications in the 1920s. Transatlantic tests took place on a variety of frequencies during the 1920s and early 1930s, the first in February 1921. Amateurs in the US were licensed to use 100W compared to just 10W in the UK. Twenty five US amateurs took part in the first tests but, disappointingly, no British amateurs received the signals.

With a special licence, the RSGB put up a special transmitting station in the works of the London Electric Supply Co at Wandsworth, where an aerial some 200ft high was erected, and 1kW of signal was used. Although signals from this station were heard in America, no two-way contacts resulted.

A second series of tests took place in December 1921 when signals were received, almost certainly by W F Burne 2KW of Sale in the first instance in the early hours of 8 December. This was despite using the smaller aerials within the limits imposed by the Post Office. The first complete message was transmitted by 1BCG, the Radio Club of America. Burne logged seven of the US stations taking part in the test and won several prizes that were being offered for successful logging of signals. Stations were using the 230m band.

In November 1923, the first two-way contact was established between the French amateur Leon Deloy 8AB and Fred Schriel at the ARRL station 1MO on 100m, despite Deloy not having permission to operate on such a low wavelength. He used a Hartley circuit with variable series condensers in both the aerial and counterpoise leads. Two French SIF 250W valves were used in parallel as oscillators with high voltage 25 cycle AC applied to their plates. Leon Delay was the President of the Radio Club de la Cote D'Azur and had made regular reception reports of hearing British signals since the summer of 1922 using his one-valve receiver.

The first British amateur to make two way contact was Jack Partridge 2KF on 8 December 1923 when he contacted A1MO, operated by Ken Warner. Partridge's transmitter used a Mullard 0-150 valve and the HT was obtained from a 1500V Mullard generator driven by a half horsepower DC motor. The receivers used at both ends were almost identical, one detector and one low frequency stage. A1MO was using 400W input and 2KF said his aerial current was only 1.8A compared with Warner's 2.5A. In a letter describing his historic contact, Partridge wrote, "At 0545GMT on December 8, A1MO first received 2KF and gave me the OK signal, wishing me good morning. He then opened up by saying "Some more amateur radio history in the making – this is the first two-way working with Great Britain. Here Warner of QST. QRA?" Contact was maintained until 0827GMT when I heard Warner say "Going now OM. Very QRZ. This is the end of a wonderful night. Goodbye"."

On 16 December 1923 the first contact between Canadian A W Greig 1BQ and Ernest J Simmonds 2OD took place on 116m using just 30W. The letter Greig wrote confirming the contact detailed his equipment, "I was using three five watt tubes when I first worked you but now have two fifty watters sitting very majestically on the shelf... The circuit used here now is a modified Hartley a la Reinartz... I use at present 1250 volts of rectified AC on the plates of the fifties and am getting exactly 6 amps radiation... My antenna is a six

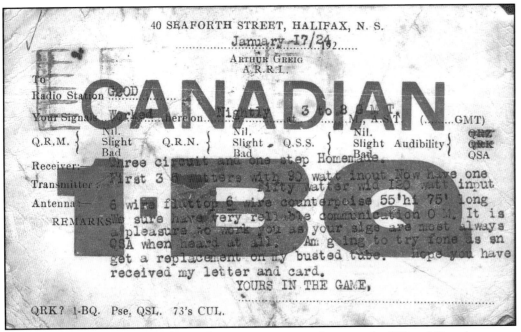

The QSL card confirming the first two-way contact between the UK and Canada.

2. Early Experiments, Famous Firsts & Pioneers

The station of Ernest Simmonds 2OD.

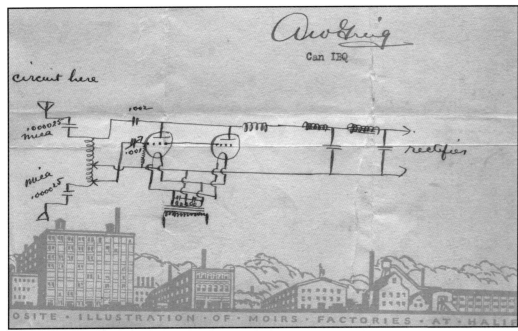

The circuit used at the station of A W Greig 1BQ from his letter marking the historic contact.

wire flat top on fourteen feet spreaders and is fifty five feet above ground and is seventy five long. The counterpoise of six wires on twenty foot spreaders directly below makes a nicely balanced system bringing the nodal point exactly at filament clip on ATI. For reception I was using a tuned radio frequency but have gone back to my old three coil regenerative as it is sure hard to beat a well designed and carefully handled regenerative set."

Cecil Goyder with the school lab radio station.

Prince of Wales visited the school and Goyder was presented to the Prince.

2. Early Experiments, Famous Firsts & Pioneers

At just 18 years of age, Cecil Goyder, using the Mill Hill School callsign G2SZ, made the first two-way contact between England and New Zealand. It certainly caught the imagination of the press at the time. Cecil was a former pupil at Mill Hill School and was studying electrical engineering at the nearby City & Guilds Engineering College. He had been allowed to use the radio station in the school laboratory.

For some months, amateurs in Europe and the USA had been logging signals on wavelengths below 100m from Australia and New Zealand. It was just a matter of time before the first British two-way contact took place. At 4am on 19 October 1924, Cecil was at the radio station in the school, a new installation had been completed overnight and he had gone to the lab to test it. He was using a two-valve set (one detector and one low frequency valve). His aerial was 40ft high but only electrically 15ft off the school roof with a counterpoise earth underneath. He was soon hearing signals from New Zealand. Around 6am he heard Z4AA calling CQ, so he went back to him tuning as close as he could to Z4AA's signal. Immediately, back came Frank Bell Z4AA reporting the signal as FB (fine business). During the conversation, Cecil learnt that Frank was located at Waihemo near Dunedin, some 12,450 miles away, and that he was a sheep farmer. Cecil asked Frank to send a cablegram to the school to prove the contact – it arrived with the headmaster the next day. Frank also sent a telegram to the Radio Society of Great Britain acknowledging the contact. In February 1924, the Prince of Wales visited the school to open the new science buildings. During a tour of the new facilities, Goyder was presented to the Prince.

The following year, he made further history by completing two-way communication with the MacMillan Expedition to the North Pole. In 1927, he became the first Briton to communicate with the Antarctic when he communicated with a whaling ship 250 miles inside the Circle.

Cecil went on to enjoy a long career in radio communications. After getting his BSc at London, he worked for a time in Paris for Standard Telephones and Cables. In 1935 he joined the BBC's technical research department concentrating on short wave work and the development of Empire broadcasting. In 1936, the Indian Government asked the BBC to "lend them a man to build the Indian broadcasting system" and Goyder undertook the task. In the next ten years he built some 15 stations and his work proved highly successful. He later worked at the UN as its first officer in charge of communication. The RSGB presented Cecil with the ROTAB trophy in 1926. He was awarded the CBE in 1946.

Saturday 19 October 1974 marked the 50th anniversary of that first radio contact between the UK and New Zealand. To mark the occasion, GB2SZ was activated by the Cray Valley Radio Society on behalf of the RSGB from the station of G2MI in Bromley. A great deal of assistance was given by G8PO, G6XN, G6DW, G3XRX, VK3AD (ex G6TM) and ZL4BX. Members of the Otago Branch of the New Zealand Association of Radio Transmitters (NZART) assisted Frank Bell in putting ZL4AA on the air and contacts were made with GB2SZ on both 80 and 40m. GB2SZ also made contact on 20m SSB with ZL4AV, ZC4BX, ZL3GS, ZLSRB, VK3AD and VK43S. VK3AD was a friend of Goyders in 1924 and often operated the same station.

Cecil Goyder sent the message, "Warmest greetings and best wishes to Frank Bell on the 50th anniversary of our first QSO England to New Zealand. I am sure that you join me in greeting and appreciation of the amateur fraternity who have recalled this memorable occasion - Cecil Goyder, ex-G2SZ".

The first contact between Australia and the UK took place between Max Howden 3BQ in Melbourne, Australia and E J Simmonds G2OD on the 95m band on 13 November 1924 at 0650GMT. Simmonds was using a 1-valve transmitter running around 150W, the aerial was 40ft high and 76ft

Centenary

The station of Ernest Simmonds 2OD in 1925. On the left you can see the 20m transmitter, on the right you can see the 90m master oscillator set with master driver valve on the panel in a carrier by the window.

long in the form of a wire cage. The HT required for the set came from the mains transformed up to 1500V. They continued to exchange messages until 0715GMT when the signals faded. Apparently, according to his letter, the press gave Howden "no peace at all". He speculated that this was because Amalgamated Wireless (Asia) Ltd were supposed to have established a commercial Australian/UK link some two years previously and had not been able to do so. In Australia, Howden used a 50ft wire cage aerial suspended with hemp rope between two 80ft masts around 100ft apart, with the lead in from the middle. This was connected to the set with inch x 1/16 copper ribbon. The counterpoise was 100 ft long. The transmitting tube was a Philips Z4 with 1400 to 1500V at about 100mA on the plate. The aerial current was about 1A.

A few days later, Simmonds was in contact with Charles Maclurcan, President of the Wireless Institute of Australia. He took the opportunity to send a message to the King, "To His Majesty the King, greetings from Australian radio experimenters, Maclurcan, President, Wireless Institute". Simmonds duly passed on the message to the palace and received a letter saying, "Should communication with Australia be re-established on this system in the near future, will you kindly convey to Mr Maclurcan an expression of his Majesty's thanks".

Simmonds went on to achieve many other 'firsts' with Mexican station 1B in January 1925, Argentinian CB8 in February 1925 and a 20m contact with AF7 in Buenos Aires in September of that year. He frequently lectured at Society meetings and it was at one of his early lectures that he described and demonstrated the principals of the super-heterodyne for short wave reception. Later in his amateur life he concentrated on VHF matters playing a prominent part in the 56MHz Field Days prior to WWII.

It was in February 1925 that Gerald Marcuse G2NM received a message from Dr Hamilton Rice on his Amazon

Gerald Marcuse G2NM's station at Sonning.

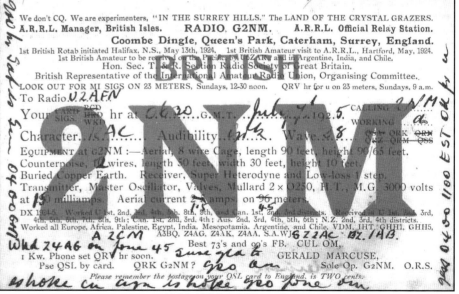
1925 G2NM QSL card.

Expedition. This was passed to the Royal Geographical Society, who requested that G2NM should send messages back to the expedition.

The field station wrote to Marcuse to thank him for his handling of the messages and said, "Good communications not unlike good conversation takes two and I must say you did not fall down at your end but demonstrated that you not only have a first class station but understand thoroughly its operation." The conditions under which the field station operated made it all the more remarkable that any message got through. The station had been off the air between 2 and 20 February when the exploration party was advancing up the river. Apparently, some days it took from dawn to dusk to travel barely a mile. The operator related that, "Three times the canoe in which my radio apparatus was loaded came near being swamped in the whirlpools and rapids and on two occasions my radio equipment was thoroughly water soaked only to be placed in the sun to dry and the journey continued". The field station, which was donated to the Royal Geographical Society, was a 25W VT transmitter using two RCA UV230A 50W valves with a small generator delivering 500V to the plate – not the rated 1000V.

In August 1925 Mr Bageley G2NB, Coventry, established the first two-way communication with Bermuda on the 45m band. His transmitter used Marconi T250 valves that had been specifically made for the work. The HT supply was from a transformer through two MR4 rectifying valves with a consequent tone that caused his signals to be described as DC. The station in Bermuda, callsign BER, was working on 35.2m.

The same month Gerald Marcuse G2NM, Caterham, created another record by carrying out the first two-way telephony contact with Tasmania. He also used Marconi T250 valves in the transmitter and used a DC high voltage generator to supply the HT to the anodes of the modulator and oscillator valves. The Tasmanian station was operated by Mr Brookes and he reported that the contact took place at 0725GMT.

G2NM also spoke regularly to the *USS Seattle* when at anchor in Wellington Harbour with both stations in daylight. In fact, the Commander of the *Seattle*, at first, refused to believe that he was listening to an experimenter in England.

It was after one of these conversations that Marcuse heard the ship exchange messages with the ships of the MacMillan Arctic Expedition in Greenland.

Later that month, G2NM added another first, te-

E J Ernest Simmonds G2OD

E J Simmonds G2OD, of Gerrards Cross, a Bank Manager by profession, was amongst the most active experimenters in the early 1920s. In addition to having a number of successful long distance contacts, his equipment was constantly being developed and he wrote many articles about it. Like most of his contemporaries, Simmonds started as a listener. His receiving station used a large aerial tuning inductance. By 1922 he was using a superheterodyne receiver, it did not employ a beat frequency oscillator, instead it relied on making the middle stage of the IF amplifier oscillate by adjusting its bias. On the transmitting side, Simmonds was also a prolific experimenter. He soon graduated from the single valve transmitter to a master oscillator arrangement, using valves such as the Mullard O/150 and O/250. He was fortunate enough to have an AC supply available and used a home constructed rotary rectifier. Later he changed to an accumulator HT supply in an attempt to achieve the much sought after "pure DC note". For an aerial he used a simple cage type with counterpoise.

2. Early Experiments, Famous Firsts & Pioneers

Barbara Dunn G6YL (front left) and Nell Corry G2YL (right) seen here at the 1948 VHF Convention in London

lephony during daylight with Mosul in Iraq. Tests were first carried out at night times and the signals recorded were of such strength they decided to attempt the daylight tests. The distance was 2050 miles, completed with 400W on the 45m band.

THE FIRST LADIES OF AMATEUR RADIO

Barbara Dunn G6YL was the first lady to obtain an amateur licence in 1927. She was particularly interested in the science of wireless transmissions and her early log books show extensive reports about the signals she received from amateurs, aeronautical, broadcasting, experimental and maritime stations. Entries were accompanied by the prevailing atmospheric and weather conditions and any appropriate press cuttings regarding the more famous loggings (see the Southern Cross story on page 40). Barbara received a letter from the Post Office saying that she would be granted a transmitting licence for CW on 23, 24, 90 and 150-200m, but that was only provided after she was able to show her Morse competency. On 19 August she went to Chelmsford Post Office and "did my Morse test... received at over 13 words per minutes (sometimes quicker) and sent at 14 words per minutes. Examiner said I'd done 'very well

Centenary

Nell Corry G2YL with her homebrew transmitter and receiver in February 1934.

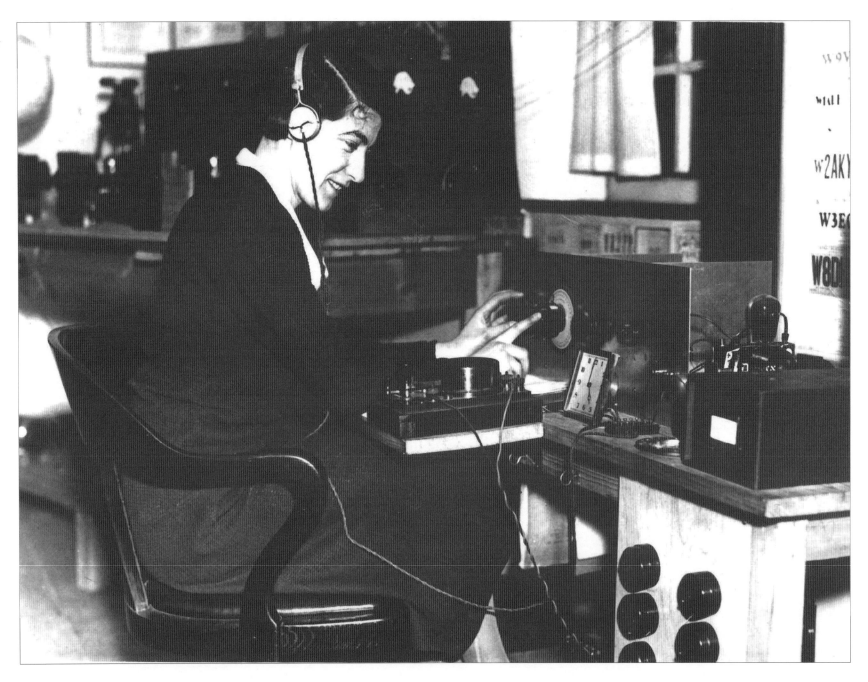

indeed'." She had been given a piece of text from *The Times* to send and was stopped because she was adding in all the punctuation when she only needed to include basic punctuation. The examiner finally sent at 25 wpm to see how she got on – she copied everything correctly. Not bad for someone who was self taught! She received her licence on 1 September and made her first call to GI6YW using 45m on 21 November. She became a very well known CW operator with excellent sending and very accurate receiving qualities. Her station was CW only as, "I don't like phone", something she continued throughout her life. Her first transmitter had an LS5 in a Hartley circuit, with about 350V derived from a Mortley rotary converter running off a 6V car accumulator feeding a half-wave VF Hertz.

In 1930 she won the 1930 Committee Cup for outstanding work in the first series of 1.7MHz tests. She became the first woman to hold an RSGB trophy.

John R 'Reg' Witty G5WQ was Barbara's brother and built much of her equipment. Both G5WQ and G6YL helped in WWII as Voluntary Interceptors in the Radio Secret Service, where their role was to listen for German transmissions.

Nell Corry G2YL became the second lady to be licensed in 1932. She very quickly took a lively interest in many technical aspects of amateur radio. During the 1930s Denis W. Heightman G6DH became fascinated with operating on the 10 and 5m bands. Very few British amateurs used these bands where equipment had to be home-built. In early 1937 he reported on his 28MHz monitoring in the *T & R Bulletin*. "A strange phenomenon, first observed by the writer in late 1935, was the appearance, at irregular times, of a radiation which took the form of a smooth hissing sound, when listened to on a receiver. It was pointed out by Nell Corry G2YL that on the days when hiss was heard that there had frequently been fade-outs or poor conditions on the high frequencies.... The phenomenon apparently originates on the sun, since it has only been heard during daylight, and it has been suggested that it is caused by a stream of particles shot off from the sun during abnormal activity."

Members of the Propagation Group received cosmic data regularly for a year making it possible to study propagation in a way not previously done. The *T & R Bulletin* published magnetic conditions, sunspot information, solar prominence activity, times of observed chromospheric eruptions from the sun and ionospheric measurements. Using 28 to 30MHz, each member was asked to assess a figure from 0 to 5 to each of four types of signal (East, 1000+ miles to the east, eg India; South, 1000+ miles to the south, eg Africa and South America; West, 1000+ miles to the west; Europe, any direction under 1000 miles). The stations involved in this research were Nell Corry G2YL, Denis Heightman G6DH, W R Emery G6QZ, Francis Garnett G6XL, G8CO, G8MH, G8SA, BRS25, BRS3003 and BRS3179.

Nell coordinated the reports and kept detailed logs for the 28MHz amateurs. She wrote, "This takes the form of a sudden increase of background noise over the whole band, and usually lasts from ten seconds to two or three minutes.... It has been heard on...56Mc., as well as 7, 14, and 28Mc., but it is usually strongest around 28Mc. It is definitely due to some form of intense solar activity, and on at least seven occasions has been heard on the same day as a Dillinger Fade-out." Nell's compilations were analysed and over the 1936-1939 period it showed that 24 amateurs in America, Britain, India and South Africa reported the hiss phenomenon on 107 different days, on frequencies ranging from 7 to 56MHz.

Nell made history in 1935 by working a British Empire station in each of the five continents on 28MHz during a period of 6 hours and 20 minutes, all using her home made equipment.

Constance Hall G8LY obtained her full amateur radio transmitting licence in 1936 at the age of 24, the third lady to hold a licence. She constructed much of her own transmitting and receiving equipment that she set up and

A D Gay G6NF, RSGB President 1941-1943

After the Paris Congress in 1925, a dinner was held in the Waldorf Hotel, London for Hiram Maxim; Alfred Gay G6NF was one of the amateurs that attended that evening. He was also involved in the tests carried out from Crystal Palace in May 1933 on 56-60MHz. As one of the founding members of the Technical Committee, created to maintain the high standards of the technical publishing of the RSGB, he oversaw many of the Society's technical books. G6NF served on the Council from 1930 holding the offices of Honorary Treasurer and Honorary Vice President, he was made President in 1941. He had made outstanding contributions to the work of the Society by undertaking the onerous duties of Calibration Manager (1929-1939) and had achieved national recognition in the specialised field of frequency measurement. He contributed to the *T & R Bulletin* and to *The Amateur Radio Handbook* and was in frequent demand as a lecturer. He won the Braaten Trophy three times having led the British Isles entrants in the Annual ARRL DX Telegraphy Contest for the years 1937, 1938 and 1939. He was an active VHF enthusiast in the 12 years leading up to his Presidency. He gave his presidential address on Frequency Standards in London during the Blitz. He was made an Honorary Member of the Society in 1944 to mark the appreciation of the Council for steering the Society so successfully through the first three years of WWII. Professionally he was the Chief Chemist for Schweppes.

E Dawson Ostermeyer G5AR

Ernest Dawson Ostermeyer was one of the first people to join the Wireless Society of London after WWI but did not take an active role in the work of the Society until 1928. He represented, with others, British radio amateurs at the first congress of the International Amateur Radio Union in Paris in 1925, becoming RSGB Honorary Treasurer in 1929 and RSGB President in 1937. He maintained a very efficient station at his home in Woodford. He presented the Society with a trophy for an annual award for the best home constructed equipment.

E Dawson Ostermeyer G5AR and his early radio equipment.

operated from her bedroom. Without the benefits of mains electricity, this was quite a task. Her parents could not understand her interest in the technical aspects of wireless but accepted the situation. Holes were drilled into window frames to accept the lead-in cables of the many aerial wires strung around the garden. Constance and Nell were great friends until Nell's death in the 1970s. Before the outbreak of WWII, Constance became interested in the new science of Very High Frequency transmissions (VHF) and was one of the early pioneers transmitting on the 56MHz band.

Prior to the outbreak of WWII, the family moved from North Waltham to Lee on Solent where she continued her wireless interests. Keen on using Morse code and an extremely competent operator, Constance's voice and 'Morse fist' were well known on the amateur bands. During the early days of WWII, her technical and radio operating skills were recognised and she was recruited as a Voluntary Interceptor. During this period, Constance conducted her listening activities from a former garage in the garden of her home. Family and visitors alike were barred entry. She never spoke openly about the period and received the Defence Medal for her work.

At the outbreak of WWII, both Nell Corry G2YL and Constance Hall G8LY contributed to Arthur Milne G2MI's The Month 'off' the Air column. Commenting on the 56MHz band, Constance said;

"Hang up your headphones on the old shack wall
And cuss, cuss, cuss
Hang up your headphones on the old shack wall
But do not make a fuss
What's the use of listening
It hardly is worth while, so
Hang up your headphones on the old shack wall
But smile, smile, smile

It was in the sense of this lament based on that old favourite 'pack up your troubles' that the majority of 56MHz and UHF enthusiasts reacted when the news came through that their licences had been determined.... What should a keen UHF ham do during these long 'blackout' evenings? There can be but one reply - if you have access to your shack, set about improving your receiving gear and other non-transmitting apparatus."

SCIENTIFIC TESTS

In June 1927, RSGB Members were invited to carry out tests on the short wave bands in conjunction with the Board of Scientific and Industrial Research during the solar eclipse. Certain stations were appointed to transmit on the 23, 44, 46, 90 and 100m bands (23, 44 and 46 for CW and 90 and 100 for modulated carrier wave). These tests were expected to produce information on the height of the Heaviside layer during the eclipse. The Heaviside layer is known as the E layer today, a layer of ionised gas at a height of around 90-150km that reflects radio waves back to earth.

Ralph Royle G2WJ was involved in a radio experiment, organised by the Institution of Electrical Engineers (IEE), which centred on the eclipse. He was to transmit a message on the 46m band on the two days before the eclipse as well as the day of the eclipse. On the two days following the eclipse he was asked to transmit "any piece of newspaper matter handy". In the supplied message, listeners were asked to write down the whole of the transmission from 04.40 to 05.40GMT each morning until 1 July. They were to indicate the strength of signal by the size of the written character. Listeners were asked to not adjust their sets during the broadcast, especially during the eclipse itself. They were asked to send their reports to the IEE in London. Ralph sent this message repeatedly during the 1 hour transmission, being careful not to miss any of it out. The transmission

2. Early Experiments, Famous Firsts & Pioneers

Acknowledgement of Ralph Royle's part in the research undertaken during the 1927 eclipse.

> TEL: VICTORIA 8228.
> VICTORIA 7290.
> 9, Buckingham Street,
> Westminster, S.W.1.
>
> 13. 7. 27
>
> Thank you very much for your interesting eclipse record. Observations are still coming in & the results will be sent to you later
>
> W. Eccles.

Ralph Royle G2WJ at his station.

was 12 lines of three letter groups, for example, osu asy itu iwu aby ugy amo avo ofe upy.

Others involved were G5YG in Glasgow on the 100m band, G2NM in Surrey on the 90m Band, G2OD in Gerrards Cross on the 32m band, G6WW in Leicester on the 44m band and G6IZ in Aberdeen on the 23m band.

Barbara Dunn G6YL took part in the tests between 0526 and 0717 on 29 June and logged 35 amateur and broadcast signals, mainly from the USA. Another participant was Edwin Alway stationed at RAF Heliopolis in Egypt who intercepted the codes on 29 June. He wrote an article on the tests for the Institute of Radio Engineers in 1927. He concluded, "At least locally, a solar eclipse produces a pseudo night effect, this effect beginning with he eclipse but finishing before it; that is, when the eclipse occurs in the morning."

It wasn't the only time that UK radio amateurs were involved in research during an eclipse. In 1973, during the solar eclipse of 30 June the Ionospheric Research Group of the French National Centre of Telecommunications Studies made a study of the behaviour of the equatorial ionosphere during the solar eclipse and particularly of ionisation transport phenomena.

It was requested that all radio amateurs who had contacts with stations located either in Central or South Africa during the periods 27 - 29 June 1973 and 1 - 4 July 1973 should send extracts of their logs for these periods giving date, callsign and QTH of the calling station, callsign and QTH of the called station, time (UTC) of the QSO, signal level (RST code) and peculiarities of the QSO (QSB,ORM, QRN etc). Reports from listeners of signals coming from the same area were also welcomed.

The 1912 station of H W Pope PZX (later G3HT).

Inter War Years

Gerald Marcuse G2NM

Interest in broadcasting began in Britain after the end of WWI and amateurs started experimenting with speech instead of Morse. Some, fancying themselves as entertainers, broadcasting music from gramophone records - often on Sundays - and became popular with locals who followed their broadcasts.

In 1927, Gerald Marcuse, G2NM, made successful contact with an amateur radio operator living on the island of Bermuda. This transatlantic communication grew into regular communication and the Bermudan radio operator often re-broadcast the voice communications, and sometimes radio programming, from England to other amateur radio operators in the Caribbean.

Marcuse applied to the GPO for a permit to broadcast regular programming. The official permit granted approval for Marcuse to broadcast speech and music for two hours daily, on 23 and 33m with a power limit of 1kW, for an experimental period of just six months. When the first broadcast took place, on 1 September 1927, G2NM had already spent £6,000 on the project. It is interesting to note that BBC Empire Broadcasting didn't start until 1928 after conducting their own tests from G5SW in Chelmsford in November 1927. Marcuse, G2NM, started broadcasting regularly on 11 September, going on the air every Sunday at 6am and 6pm.

Marcuse received one of the Empire broadcasts from Australia and rebroadcast it over his own short wave station and beamed it back to Australia. Many QSL cards were issued to verify the reception of these broadcasts, and later QSL cards issued by Gerald Marcuse at his amateur station G2NM acknowledged the fact that his station had broadcast the first series of Empire Broadcasts from England. The original experimental licence was extended for an additional six months.

2NM's first interest in wireless experiments dated back to 1913 but his impact on the amateur radio movement was not felt until just after WWI when he became Honorary Secretary of the Radio Transmitters' Society. After many months negotiation, the RTS finally merged with the T & R Section of the RSGB in 1925. Marcuse became Honorary Secretary of the T & R Section and took part of the decision to publish a monthly journal. In May of the same year, he

Gerald Marcuse G2NM.

received a letter from the GPO that stated that information had been received from the Japanese Telegraph Authority that signals from G2NM had been heard in Japan. The letter asked Marcuse to supply precise information regarding the power and wavelength used and concluded by saying that "the information, if furnished, will be regarded as confidential and used for technical purposes only, while if the limitations of your licence have been exceeded in the tests, steps will be taken, if possible, to amend the licence to regularise the tests." He was, very naturally, proud of that letter. In 1946 he was elected an Honorary Member in recognition of his outstanding services to amateur radio in general and to the Society in particular.

In addition to his interest in amateur radio, Marcuse was an international figure in yachting circles, having won many prizes in home and foreign waters. He held a ship-to-shore licence for the various yachts he had owned. This included *My Babe II* that was built in Littlehampton in 1939 for G2NM on which he installed ship-to-shore radio in the wheelhouse – which was much appreciated by the Navy when they took her over at the outbreak of WWII. It had only been on one sightseeing trip for the owners to Leige, via Ostend, for the unveiling of the Albert Monument by King Leopold. The craft was one of the Dunkirk Little Ships and, when the Navy offered to compensate him, G2NM assumed the craft lost. He didn't accept payment, preferring to look on the loss as his contribution to the war effort. The family learnt later that the ship had survived Dunkirk and may still be afloat today.

Gerald Marcuse G2NM the QSL card from an annual special event celebrating his part of Empire broadcasting.

The driver for G2NM's Empire broadcasting equipment.

The Marconi staff at Writtle who were responsible for building the 2MT station that took over Empire Broadcasting from 2NM. Courtesy Marconi's Wireless Telegraph Co Ltd.

Following his death in 1961, a bench and sundial were placed in the grounds of Bosham church to commemorate the life and work of G2NM.

CALLSIGNS & QSL CARDS.

In the early 1920s, callsigns could be misleading. In the UK, L E Owen had the callsign 2AB, but so did J D Jarest in Quebec, Canada; D Wilkinson in New Zealand; A V Badger in Sydney, Australia; C Quirit in the Philippines, Mr Justi in Brazil and Mr Illa in Uruguay and there were probably more. With international contacts on the rise, it made identification a little difficult. Many amateur stations added a prefix letter to their callsigns to identify their country of origin. British stations adopted G, French stations F and, to begin with, stations from the USA often used the letter U.

The British Post Office made Britain one of the first countries to adopt a recognised prefix in October 1924. Other countries soon followed. The use of ITU allocated prefixes didn't become widespread until the very end of the 1920s. In 1926, the Post Office agreed to the prefix GI for Northern Ireland, although Scotland and Wales didn't get separate prefixes until several years later.

No one seems to be quite sure who invented the QSL card – printed post cards used to confirm contacts or to acknowledge reports. It is likely that some form of acknowledgement card was used by amateurs in the US prior to WWI.

John Karlson SMUA - later SM6UA

John Karlson lived in Goteborg, Sweden and was one of the most active Swedish DXers before WWII. He was a well-known pharmacist and started as a SWL in 1925. Licensed in 1927 as SM6UA, he was a staunch believer in the powers of amateur radio to increase international goodwill. After his death in 1940, his son SM6UB planned to get back on the air after the war.

3. Inter War Years

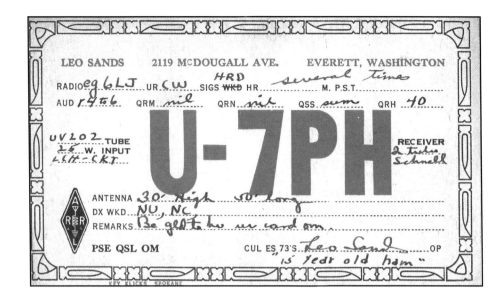
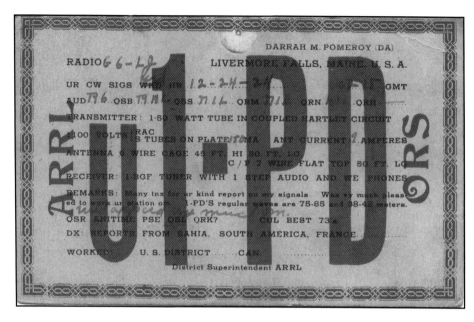
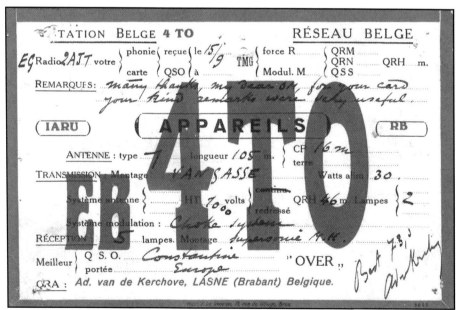

QSL cards from the 1920s, note the US cards with the U prefix.

Centenary

TECHNICAL TALKS

In November 1922, the Wireless Society of London changed its name for the last time to the Radio Society of Great Britain at a Special General Meeting. The resolution to change the name was put to the meeting by Sir Chas Bright and seconded by Mr Edward Shaughnessey OBE – it was carried unanimously. It was also suggested that the Society should consider awarding a medal each year to the amateur radio experimenter who achieved the best work 'for the Science'. This is something that continues today with the various awards that take place at the AGM each year.

At this time, some 286 transmitting licences and 6986 receiving licences had been issued.

In the early days of the Society, the meetings included talks from some of the most prominent amateurs of the day as well as eminent scientists. Major J Erskine-Murray (callsign MUX) had been a pupil of Lord Kelvin for 6 years at Glasgow University. In 1898 he became Marconi's experimental assistant and visited Marconi's house – the Haven – on a number of occasions. He was known to play the cello, accompanying Alfonso, Marconi's brother, on the violin with Marconi, himself, on the piano. During WWI he was the officer in charge of the wireless experimental section of the RAF. He was a well known speaker to many organisations such as the IEE, as well as the Society, and an author of several papers and technical books. In 1921 he spoke to the Society about 'the greatest problem in radio', where he described how to prevent unwanted signals rendering wanted signals unreadable. This was later included in his book *A Handbook of Wireless Telegraphy*.

Another famous speaker was Sir Oliver Lodge, who became President of the Society in 1925. Lodge was one of the early pioneers in the application of Hertzian waves and in 1894 gave a lecture at Oxford University on the work of

In 1925, Sir Oliver Lodge, who had been made an Honorary Member of the Society in 1914, became President. Lodge was a many-sided man; his research into radio waves and the behaviour of electrons had brought him a knighthood. He was the first to describe the mechanism of 'secondary emission' and explain its effect on the characteristics of a tetrode – an effect that is still relevant to the users of the famous 4CX250 family of valves.

Hertz where he transmitted radio signals to demonstrate their potential for communication. He was appointed as a scientific adviser to Marconi's company. He was a popular broadcaster in the early days of the BBC. In 1922, he gave the lecture that attracted the biggest audience as well as the most publicity. He spoke about his early experimental work in the production of electric-magnetic waves in space.

Other activities linked to these early meetings were visits to such places as the Experimental Stations of the Radio Communication Company in Slough and Croydon Aerodrome to see the wireless installation where both the receiving and transmitting equipment was built by Marconi's Wireless Telegraph Company.

Professor W H Eccles was a British physicist and a pioneer in the development of radio communication. In 1898, he became an assistant to Marconi, and whilst working at the Chelmsford factory he devised the first bench method of testing coherers whilst avoiding having to connect them to an aerial. In 1913 he was part of the Advisory Committee set up to give technical advice to members and to put the ideas of the Society to the Post Office. Eccles suggested that solar radiation was responsible for the observed differences in radio wave propagation during the day and night. In September 1923 he gave a lecture to the Wireless Society of London on The Amateurs Part in Wireless Development, in which he talked about the various great advances in wireless that had been initiated by the amateur. He went on to discuss the idea that "every large city and town may have its own broadcasting station some day is not so fantastic". The following year his talk to the Society was on The Importance Of The Amateur when he discussed the problems arising in a world where there was an ever increasing use of wireless by broadcast and commercial companies thereby putting pressure on the amateur.

Professor W H Eccles, RSGB President 1924.

W K (Ken) Alford 2DX (formerly TXK)

Ken Alford was interested in scientific matters early in his life and obtained an experimental wireless licence with the call TXK in 1912, aged 19. The station was a 10 inch spark transmitter. It was powered by a 35V 12A dynamo driven by a gas engine that charged a 14-cell battery for the spark coil. The aerial was a 48ft four-wire cage. Tuning was by means of a spider-web coil and a bank of nine Leyden jars. The receiver was a typical crystal type using Perikon detectors. He was one of the few stations to copy the names of the Titanic survivors. After WWI he had the callsign 2DX and was part of the group that formed the T & R Section. He was active in the transatlantic tests of 1921-1923 and after seeing the superheterodyne receiver brought over from the US by Paul Godley, he mastered the principle, published articles and helped friends such as 2NM and 2OD acquire such receivers. His first station was described in the July 1923 edition of *Modern Wireless*. The aerial was two four-wire cages 50ft long at 50ft above the ground. The transmitter made use of three R-type valves connected in parallel. Speech modulation was with by the 'direct method' in the aerial circuit or the 'grid leak' method. The operating wavelength was 195m. The power supply was provided by a gas engine built before WWI from castings; this drove a dynamo for charging accumulators and supplying the high voltage transformer. The receiver was one of the first home constructed superheterodynes. It had 10 valves, most of which were used in the long wave amplifier operating on 3,000m. By October 1924 the transmitter had been completely rebuilt and now had a master oscillator with two AT40 valves in parallel driving a Mullard 0/150 valve amplifier with 2kV on the anode. Rectification was carried out by a pair of Mullard U50 half wave rectifiers. The transmitter was tuneable between 70 and 120m.

In 1967, he still had, intact, the DET1 valve with which he had worked Australia in 1925 – the Bulletin carried an advert for the valve on December 1925 saying "the DET1 was the first dull emitter valve to communicate with Australia at an input of 66 watts". His log contained two particularly historic entries:

Aug 4 1914. POZ (Nauen) "Krieg ist erklaert gegen Frankreich und Russland" (War is declared against France and Russia). His friends and neighbours didn't believe him until the newspapers the following day.

Nov 11 1918. FL (Paris) "La guerre est fini" (War is over). They believed him this time!

Ken was on the air for 73 years before joining the ranks of silent keys.

3. Inter War Years

PATRONS

In October 1922, His Royal Highness, The Prince of Wales visited the first all-British Wireless Exhibition and Convention at the Horticultural Hall, London. In his capacity as Chief Scout for Wales, he gave a broadcast address to the Boy Scouts of Great Britain. Four days later the Society received a letter from him accepting the role of Patron, a role he continued until his accession to the throne in 1936.

To celebrate the Patron's 36th birthday, on 23 June 1930, stations around the Commonwealth (registered as British Empire Radio Union (BERU) stations) were invited to send, via amateur radio, loyal greetings to His Royal Highness. These messages were gathered at the Society's headquarters and taken to York House, one of the official residences. The Society received a letter saying, "Perhaps you would be good enough to convey to those radio relay leagues, groups and other amateurs whose greetings to the Prince you have forwarded, the enclosed message from His Royal Highness. The Prince of Wales sends you sincere thanks for your good wishes, which His Royal Highness much appreciates."

Later that year, during the Convention, it was proposed that an annual BERU day should take place. From this suggestion the first BERU Contest was inaugurated – the Commonwealth Contest as it is known today.

Early in 1952, advice was sought from Past President, Lt Col Sir Ian Fraser, CBE MP on seeking Royal Patronage once more. In May, the RSGB approached His Royal Highness, The Duke of Edinburgh, KG and he generously extended his Patronage in November 1952.

On 1 November 1952, an announcement came from Buckingham Palace with the news that 'despite having undertaken a great number of additional responsibilities, he will do his utmost to take a personal interest in the Society.' The announcement appeared in the November 1952 *RSGB Bulletin*.

His Royal Highness, The Duke of Edinburgh has visited

His Royal Highness, The Prince of Wales, the RSGB's first Patron.

Centenary

a number of events with the RSGB including special event stations, anniversary events and exhibitions. In 1988, he attended the 75th Anniversary Convention at the NEC. After touring the display of amateur equipment through the ages he exchanged a short greetings message with Windsor Castle. His opening address was carried on a special edition of GB2RS that was carried live from the NEC.

His Royal Highness, The Duke of Edinburgh, RSGB Patron since 1952.

T & R SECTION

In February 1923, the T & R Section (Transmitter and Relay Section) formed as a sub-section of the Radio Society of Great Britain, mainly with the aim of carrying out certain transatlantic tests from an experimental station erected in West London. Later that year, the British Wireless Relay League merged with the T & R Section to "ensure the continuance of its activities and give it the advantage of financial and administrative support of the larger body". The objectives of the Section were "to promote intercommunications between experimenters and thus assist them to improve their apparatus, to join hands with similar organisations overseas, to investigate the quality of the transmissions in various directions at different hours and to establish a collection of wavemeters and other useful apparatus for loan within the Section."

In June, a broadcast was made on 2LO, the first time the actual proceedings of the amateur were made public. Read by W E F Corsham 2UV, it introduced plans for radio traffic management. A chain of stations all around the UK had been formed, with the sole job of being of assistance to all amateurs within their area. A test had been set up, "to send a message round England, Scotland and Wales, originating in London, through the Eastern Counties to Newcastle, from thence to Scotland, and back through Sheffield, Shrewsbury and Bristol to Camberley and London again, on wavelengths between 150 and 200 metres." The broadcast finished with the words, "Perhaps, therefore the day when the British amateur will have encircled the World is not so very far ahead after all."

The T & R Section were asked by the Admiralty to cooperate with an experimental voyage of *HMS Yarmouth* that was on a 12-month cruise to Hong Kong and back, starting in early 1925. The purpose of the voyage was to test out short wave as a means of communication. During daylight the tests took place on the 23m (13MHz) bands and on the 24 and 46m (6MHz) bands during the hours of darkness. The Admiralty planned to carry out tests on frequencies between 6 and

The T & R Section badge.

25MHz to determine ground wave ranges and skip distances. A timetable was issued to members of the T & R Section – and a complicated one at that! Three timetables were arranged for even days and three for the odd days. It also depended on whether the Yarmouth was between Portsmouth and Malta or Malta and Hong Kong as to which of the odd or even timetables were to be used. It also depended on whether the ship was in port or not. Complicated the timetable may have been, but it must have been effective because the Admiralty stated that their Signal School Experimental Officers were able to determine that frequencies changed due to the effects of day, night and season.

In 1926, it was agreed to merge the T & R section with the main Society.

INTERNATIONAL AMATEUR RADIO UNION

Early in 1924, Hiram P. Maxim 1AW, President of the Amateur Radio Relay League (ARRL) visited Europe with the aim to "encourage international amateur radio relations". It was felt that the ARRL, representing American and Canadian amateurs, could join with other national societies to create an international organisation for the promotion of amateur radio.

In March 1924, a group of radio amateurs from Belgium, Canada, France, Great Britain, Italy, Luxembourg, Spain, Switzerland and the USA met in Paris. The RSGB were represented by Gerald Marcuse 2NM. The meeting made preliminary plans for an international organisation – the International Amateur Radio Union (IARU). Maxim came to

One of the committees, note G2NM at the back.

Delegates at the first IARU Congress.

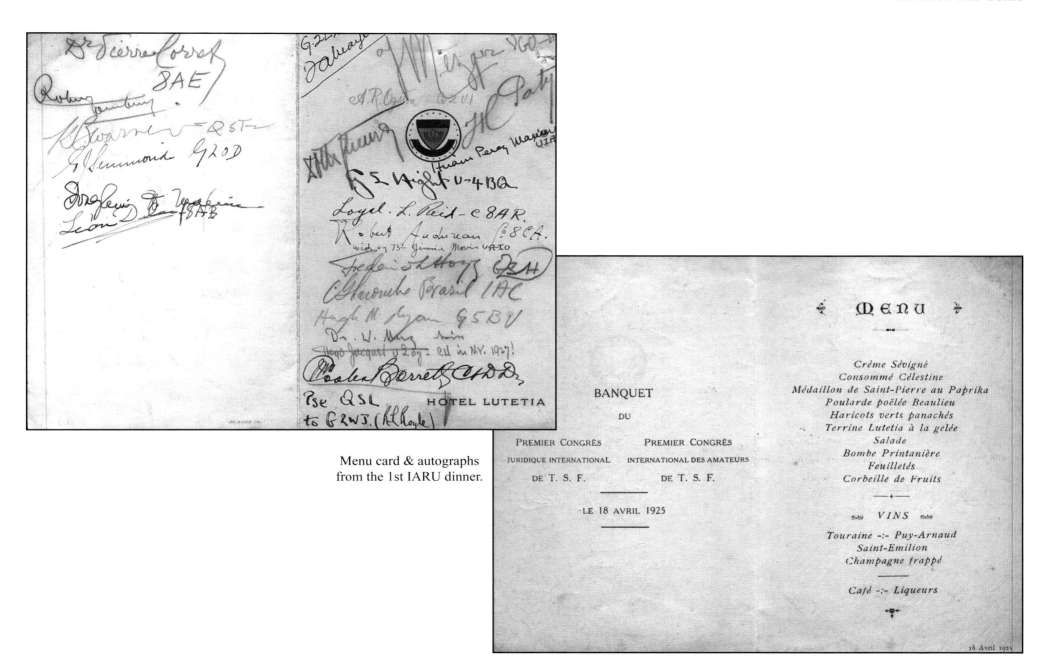

Menu card & autographs from the 1st IARU dinner.

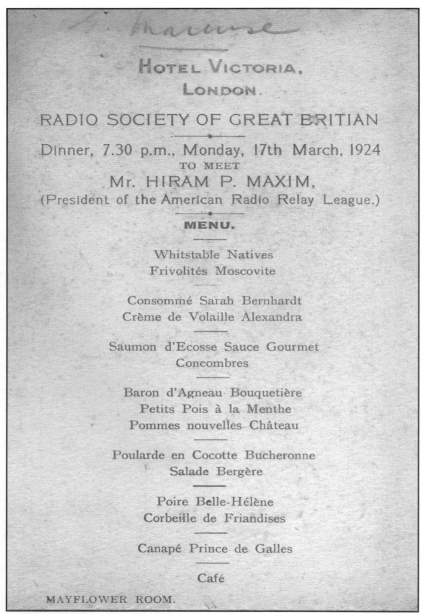

Menu card from the dinner in honour of Hiram Maxim's visit to London.

London later that month where he was the guest of the RSGB. A dinner was held in honour of the visit and was attended by many famous names in amateur radio – Hugh Pocock, Editor of *Wireless World*, Philip Coursey, RSGB Honorary Secretary, Ralph Royle 2WJ, Gerald Marcuse 2NM, both members of the T & R Section, W H Eccles, RSGB President and Maurice Child amongst many others.

A Congress was arranged in Paris in April the following year to formally create the organisation and representatives from 23 countries attended. Thirty-eight British amateurs were amongst the 250 delegates from all around the world. The RSGB were well represented on the various committees by Ralph Royle G2WJ, Hugh Ryan G5BV, Harold Bailey G2UF, Fred Hogg G2SH and Stanley Lewer G6LJ.

Probably the most important work of the Congress was the formal creation of the International Amateur Radio Union whose objective was, "the promotion and coordination of two way radio communication between the amateurs of the

Ralph Royle's delegates card.

various countries of the world; the advancement of the radio art; the representation of two-way radio communication interests in international communication conferences and the encouragement of international fraternisation".

The first International President was Hiram Maxim U1AW and Gerald Marcuse G2NM was International Vice President. Marcuse even received congratulations for his appointment from the BBC.

A British Section of the IARU was formed in late 1925. W G Dixon G5MO asked "all who are at present actively engaged on short wave work by sending me a card on the last day of each month reporting any items of interest concerning reception condition, DX worked and heard, new local short wave clubs formed; in fact any news that can be made interesting to readers of our British Section report in *QST*". *QST*, the journal for the ARRL had taken on the role of setting aside pages for IARU business.

NEW LICENCES & BADGES

A new form of transmitting licence was introduced in 1924 – much to the disapproval of both radio amateurs and the technical press. It banned all international working by British amateurs except by special authorisation. Working was limited to stations in Great Britain and Northern Ireland. Both *Wireless World* and *Wireless Weekly* immediately offered £500 to the RSGB to allow a test case to be argued in the courts. Professor W H Eccles was the RSGB President at that time and he was able to show how the new restrictions appeared to contravene the 1904 Act. Despite the meticulous planning, the arguments were never put to the test. Surprised by the strength and unity of the amateurs, the authorities wavered.

The first Society badge appeared in 1924. The original design was adopted as a result of a competition. It was a combination of two designs submitted by H W Taylor, 2QA and H H Townley. In the original badge, the head of Britannia is seen over a scroll bearing the motto *'Per Aether Is Undas'*,

The original RSGB badge that has been resurrected to create a badge to celebrate the RSGB Centenary and is a prominent part of the RSGB Centenary logo too.

The first *T & R Bulletin* Cover.

meaning Through the Ether Waves. Later, the diamond was introduced for the T & R section, no doubt based on the ARRL badge that used the graphical symbols of aerial, earth and inductance. Both badges were current for a few years. In 1926 it was agreed to merge the T & R section with the main Society and, eventually, the diamond badge became the Society's badge. Today this graphical representation is common throughout the world of amateur radio. The design of the original badge has made a reappearance for the RSGB Centenary year.

T & R BULLETIN

Until the end of 1924, *Wireless World and Radio Review* was the official journal of the RSGB. When that changed hands at the end of 1924, the Society's journal was then published in *Experimental Wireless* and the *Wireless Engineer*. In 1925, the committee of the T & R Section decided the time was right to produce a monthly publication devoted entirely to the interest of the transmitting amateur.

The first *T & R Bulletin* was just 12 pages, published in July 1925, and included five pages of advertising and the description of a single stage 23m transmitter by Ralph Royle, 2WJ. The magazine had an editorial team of three – J A J Cooper G5TR was editor, with Henry Bevan Swift G2TI and Ralph Royle G2WJ making up the team. Royle was an expert in the art of block making as his family's firm, W R Royle and Sons, were government printers. Interestingly, the first edition also makes numerous mentions of "hams", so the idea that the word ham is a relatively recent Americanism probably isn't correct.

The T & R moniker persisted as the title of the Radio Society of Great Britain's technical journal for a further 17 years or so until 1942, when the name of the publication was abbreviated to the *RSGB Bulletin* – or simply The Bull, as many amateurs nicknamed it.

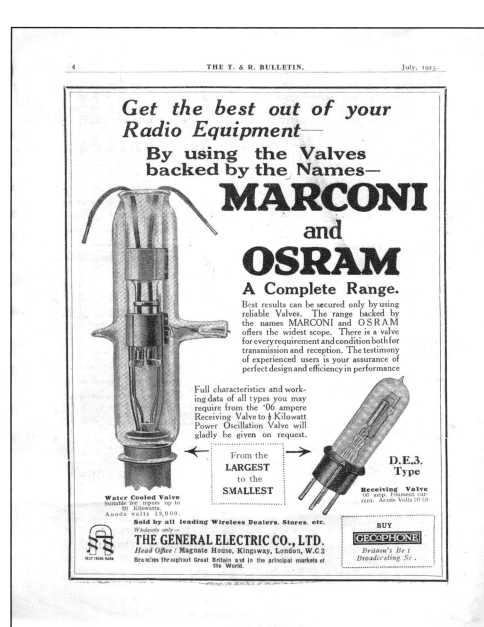

Adverts from the first *T & R Bulletin*.

In January 1968, the journal changed its name again to *Radio Communication*. The editorial that greeted this change in name reassured Members that it didn't mean a change in direction, "for the advancement of amateur radio" remained the paramount aim of the magazine. It was decided that it was a disadvantage to publish a journal under a title that means nothing to the uninitiated. The name Radio Communication was felt to convey the field covered, made it easily identifiable and reflected the role the hobby played. At the time it was noted that "Some may regret that Radio Communication does not readily lend itself to any diminutive name. It may well be that many members will continue to refer (we hope affectionately) to the Bull".

Well, in 1995, *Radio Communication* was shortened to *RadCom* and the magazine in its current format was born.

ACTING AS EXTRA SECURITY

GEC contacted the RSGB in June 1930 asking amateurs to listen out for Captain Kingsford Smith's flight across the Atlantic. The Southern Cross, the plane that he had flown from Australia to England in 13 days, would be carrying a crew of four – Captain Kingsford Smith, a 2nd pilot A Vandyke, navigator Captain J Saul and wireless operator J Stannage. The plane set off from Portmarnock Strand near Dublin, which had been chosen as it had a runway around 3 miles long. It was felt that, as the aircraft weighed somewhere between 8 and 9 tonnes fully laden, it would need that length of runway.

This was to be the first time that a full wireless station was carried on a transatlantic flight. The plane carried a 3-valve receiving set that could be tuned from 28 to 2000m,

Some examples of other issues of *RSGB Bulletin*, *Radio Communication* and *RadCom*

3. Inter War Years

Captain Kingsford Smith

CAPT. KINGSFORD-SMITH'S TRANSATLANTIC FLIGHT.

We are in receipt of your kind offer to co-operate with us in connection with the Transatlantic Flight of Capt. Kingsford-Smith.

We will telegraph you as soon as he takes off from Ireland which will be when the weather is favourable.

Acknowledgement of participation in the monitoring of the flight.

a short wave transmitter operating at around 33m and running 250W, as well as another transmitter operating on 600m. The Southern Cross used the callsign VM2AB.

The plan was to transmit continuously on 33m and owners of short wave sets were asked to track the progress of the Southern Cross. G6PP was one of the amateurs who volunteered to monitor the aircraft and his log shows that the first transmission was received on 24 June at 0630BST. The message was just 'CQ DE VMZAB'. Mr Pilpel (G6PP) noted in the log that the signal was "QRK R7-8, QSB to R4 and some QSX. Tone rough rac". The aircraft went on to work the New York Times station, WHD, passing message No 3. He continued to monitor the progress of the aircraft until 0220BST on 25 June.

Some who listened struggled because of the speed the wireless operator was sending; it varied between 10 and 25 words a minute. Barbara Dunn G6YL spent some 26 hours at her station noting all the communications from the aircraft. In her report she declared, "it was worth it.... wasn't it thrilling!". She sent her report to GEC and the coordinator, Mr Donisthorpe, sent her name to the Daily Mail who interviewed Barbara. She said, "our phone fairly buzzed with newspaper reporters wanting an interview with me. Help!! ... In the afternoon I escaped them all by going out to play tennis until late evening." She finished by saying, "I certainly do think our help with this flight ought to make the PMG see we are worth our salt".

Ernest Simmonds G2OD reported signals as R6 at 0900 on the 24th dropping to R4 at 1130 and R2 by 1420. The plane's signals were always recognisable through the noise by the "distinctive TT quality".

Following the successful Southern Cross monitoring exercise, a letter arrived at the Society's HQ in August 1930 requesting help in listening for signals from a round-the-world flight. To begin with, Alfred Graham & Co Ltd wouldn't even reveal who was flying or where. They just said that the aeroplane would be fitted with a special short wave transmitter operating around 40m. The RSGB met with representatives of Alfred Graham & Co and learned more about the flight.

The flight was being undertaken by the Honourable Mrs Victor Bruce, flying a Gipsy Bluebird from Croydon to Tokyo, a journey that was due to last 15 days. Mrs Bruce was a British record-breaking driver, speedboat racer and pilot.

The aircraft, callsign G ABDS, was fitted with a short wave transmitter that operated on a fixed wavelength of 35m and, according to the letter from Alfred Graham & Co Ltd, had around 60W to the anode of the oscillator. It was set up to transmit at least once an hour and they wanted amateurs to listen for the signals. They had devised a set of codes that would precede every signal. G ABDS OK meant the aircraft was in normal flight, G ABDS LG was making a landing, G ABDS MP was flying but engine or aircraft poor, G ABDS FA meant making a forced landing ashore and G ABDS SOS was making a dangerous descent, possibly in the water, and requires immediate assistance.

A 50W Graham Aircraft Type GA3 short wave transmitter and receiver working on 40m, using the callsign G5GL, was set up at the Alfred Graham works as the control centre. H J Powditch G5VL was the RSGB contact and he coordinated notifying the membership both at home and abroad and all the loggings.

The RSGB passed the information to Members around the world asking them to listen for the aircraft's signals. Starting on 19 September, the flight actually took 25 flying days and covered 10,330 miles. Mrs Bruce was the first person to fly from England to Japan and the first to fly across the Yellow Sea. Her trip lasted five months as she sailed from Tokyo to Vancouver, re-assembled the plane and flew to New York. From there she sailed to Le Havre, then flew back to Lympne Airport – a total of 19,000 miles of flight.

STEPPING INTO THE BREACH

In August 1932 a news item appeared in the *Daily Sketch*, a British national tabloid newspaper, relating the RSGB's work at government level to secure amateur frequencies. An incident had occurred in Cyprus when Government wireless communication failed, and an RSGB Member packed up his set, flew to Cyprus and set up a station there so that communication with the outside world was resumed immediately. Consequently, the Navy were asking members of the RSGB to enrol in the Royal Naval Wireless Auxiliary Reserves. To quote the Admiralty, "the work of the Radio Society, so valuable to peace, will be even more valuable in time of war". It was a definite indication from the authorities that they valued the resources of amateur radio.

NATIONAL FIELD DAY

The first National Field Day took place in June 1933. The original purpose was to demonstrate that low power portable stations, set up at short notice out of doors, were capable of maintaining reliable communications with other low power portable stations in different parts of the British Isles. The RSGB wanted to show that "if the necessity arose, the amateur radio movement in the UK could place into operation an emergency network of stations at short notice".

Each District that took part ran two stations, one was on 20 and 40m, the second station was on 80 and 160m. The first event was won by West London, District 15, with a score of 364 points, although as they attached their aerials to the masts of an old commercial station, several other entrants complained about this tactic. Viewed purely from a competitive angle, their methods were probably open to criticism but, as the event was organised as a test of portable equipment, they were probably justified in using the most advantageous aerial system available. The rules for the event were changed before the second event took place.

Thirty four stations took part – 25 in England, 5 in Scotland, 2 in Wales and 2 in Northern Ireland. The stations must have thought long and hard about where to set up their portable stations and chose locations on elevated ground. To get to their chosen sites, amateurs used just about every mode of transport from cars to bikes and caravan to donkey cart! As is typical of many, many NFDs – it rained, heavily.

Scottish District A station G5XQ with G5XQ operating. Asleep on the camp bed is G2RS.

Scottish District A station camp cooks G6ZV and G6VU.

Interestingly, VU2FP sent in a log of the stations he heard at Kailana in India. These included G2OP, G6GF, G6WL, G2NH and G5FV – the last station being the best signal he heard. Lt Beaumont, operator of the Indian station, did try calling all the participants but was, he said, "beaten in most cases by YI6HT". NFD has continued since then, only being halted by WWII. The first post-war field day took place in June 1947 and was won by Southgate Amateur Radio Club.

RECEPTION ON A TRAIN

July 1924 saw an experiment carried out by members of the RSGB to test the practicality of maintaining radio communication between a train travelling at high speed and fixed stations along the route between King's Cross and Newcastle. A secondary experiment was to test the reception of broadcast signals. In cooperation with the London and North Eastern Railway, the RSGB set up a short wave transmitter and receiver in a luggage van attached to the 7.30 Scotch Express leaving London's King's Cross. The aerial was entirely within the coach. They used the callsign 6ZZ and signals on 182.5m. Two-way contact was established with 2WD in Bedford, 5DR in Sheffield, 2DR in Shipley and 5MO in Newcastle upon Tyne. For broadcast reception they used a MHBR4 receiver (with one HF, two detector and one LF valves). 2LO was heard at full strength some 61.5 miles away, at 83 miles the signals were described as 'clear' and at 97 miles the music was audible but the speech only just intelligible. The Birmingham broadcast station 5IT was also monitored on the journey.

Following the tests it was agreed that it would be feasible to receive and transmit on the short wave bands and to receive the various BBC stations en route.

The Scottish Distinct B station aerials.

The winning station, G6YK, scored 283 points with 14 American stations, 2 from New Zealand and 1 on Martinique in the log. The other station in the District was G6WN with 81 points and their best DX from their 30 contacts was UO7OA.

Second place went to Scotland A where it was G5XQ and G6WL who obtained 121 and 236 points respectively. G5XQ worked 45 stations with the best DX being U2RE and G6WL had 42 QSOs with six US stations amongst their best DX.

Communication by radio from a train, Ben Hesketh 2PG (left) and Leslie McMichael 2FG (right).

Douglas Walters G5CV with the 56MHz gear used for the first two-way tests between light aircraft on 18 June 1933.

VHF AIRCRAFT COMMUNICATIONS TESTS

A VHF experiment started in May 1933 that proved many of the scientific experts of the day very wrong about propagation. Scientists were adamant that 'transmission by ultra short waves over more than a few miles was impossible'. Tests on the 5m band (56-60MHz) were conducted between the North Tower at Crystal Palace and a de Havilland Puss Moth aircraft chartered by the *Daily Herald* newspaper. L H Thomas G6QB persuaded the Crystal Palace management to lend him the North Tower for this experiment. A transmitter and receiver were rigged up and several local amateurs went out in their cars listening for the signals. The transmitter used two B12s in a push-pull TP-RG circuit, the modulator used two 211-Es in parallel and the microphone was fed directly into the grid circuit through a high ratio modulation transformer with no further amplification. A 1kHz tone was generated by a small oscillator run from a 100V HF battery.

The aerial was a 2.5m length of petrol piping hanging vertically from a stand-off insulator on the end of a nine foot length of battening that was poked out fishing rod fashion from the gallery of the tower. Two receivers were placed in the aircraft, both were superheterodynes. One used a split Colpitts circuit and had no LF, the other was the conventional Schnell/Reinartz type.

The first contact from Crystal Palace was with G5UY in Tottenham. Throughout the tests they made contact with G6KP in Brockley, G6UH in Limpsfield and G5XH in Croydon amongst others. The furthest QSO was with G5CV at just over 40 miles.

Douglas Walters G5CV, radio correspondent for the newspaper, used his homebrew equipment to pick up the 5m signals from Crystal Palace in the plane, setting a VHF record of 130 miles. Flying over London at 3,000ft, about a dozen 56MHz stations were heard. Those in the plane felt that, had they had more fuel, they would have been able to carry on to around 2,450 miles and still been able to hear G6QB because he was R9 at 130 miles when they finished the flight. A ground station, G6PL at Hollin Bank, Yorkshire also heard the Crystal Palace signals, although only at R2 – R4. Another experimenter, G2NH, heard the Crystal Palace signals on a kite aerial at South Harting.

A month later, also using the 5MHz band, George Jessop G6JP and Douglas Walters G5CV experimented with the first two way signals with an aircraft. G5CV was in a De Havilland Dragon-Moth chartered by the *Daily Herald* and G6JP was in a similar aircraft chartered by *Popular Wireless*. They wanted to establish two-way communication between the aircraft using their homebrew 5m equipment. G5CV's equipment was fitted into one plane and G6JP's in the other. Several seats had to be removed from these normally 6-seater planes to make room for the equipment. Both setups were similar, conventional push-pull circuits with an untuned grid coil. Two volt receiving valves were used as oscillators and modulators; the former were Osram P2s and the latter were two PT2s in parallel. The microphone was connected via the microphone transformer direct to the grids of the pentodes without any extra stages of amplification, and it was possible to increase the depth of modulation until the corner was almost entirely suppressed. The power supply was 200V from Hellesen super capacity batteries that had been supplied specially for the experiments. Maximum power used was 7W but this fell to about 4 or 5W towards the end of the two hour flight. The aerials were half wave current-fed type, twisted flex being used for the quarter wave feeders. The horizontal portion was slung inside the cabin about 6 inches from the roof. The receivers were of the conventional 3-valve super regenerative type with one LF stage.

There was tremendous QRM because the ignition leads and plugs were not screened. Both stations in the planes spoke to each other and to G2JV in Harrow as well as G6YK and G6NF. Walters G5CV kept up a running commentary whilst his plane was landing, describing the country he was passing over and the altitude. Both G6YK and G6NF reported hearing the transmission until the plane almost landed, despite the

Handing to the pilot of the glider the leather case containing the receiver, batteries and aerial used in the tests.

power having dropped to only 4W as the dry batteries were partially exhausted.

G5CV concluded that the results were encouraging as they clearly demonstrated the advantages of 56MHz working for reliable phone communications between planes and from plane to ground.

The following year, G6JP experimented with 5m communications involving gliders. In the course of the flight, instructions were received from the ground transmitter, much to the delight of the test pilot.

At the Dagenham Town Fair, Doug Wheele G3AKJ (builder of the TV camera) persuades his interviewees to see themselves on the screen of the 12 inch studio console monitor. R Oakley, who built the telestill scanner, is operating the camera. Courtesy Stenson, BATC.

EVEREST EXPEDITION

R F Loomes G6RL and N E Read G6US designed and built the portable equipment that was used during a climbing expedition to Mount Everest in early 1933. The wireless sets were required for two reasons; communication between the camps on the mountain and reception of weather bulletins send from Darjeeling some 300 miles away. The two men only had three weeks to design and build this equipment as it was a last minute decision. The climbers had felt that any equipment would be too bulky. They used an ex-Army portable trench transmitter modified for self-excited continuous wave Morse transmission on the 60 to 120m bands. The transmitter measured just 9 x 9 x 11 inches and was housed in a waterproof cover. Spare valves and accessories were in an oiled teak case lined with felt, something they hoped would cope with the 21,000ft conditions. To avoid the batteries freezing, they used inert cells for the low tension current and an ex-Army hand generator for the HT. The aerial, mast and all the equipment weighed just 40lb and could be carried by one man.

Taken off screen at the Radio Show 1955 through G2WJ/T camera and equipment.

AMATEUR TELEVISION

The Post Office agreed to issue amateur television licences in September 1934, but only to those who could justify their applications. The UK was the first administration to assign frequencies to amateurs for television experiments – 30-32MHz for pictures and 28-30MHz for sound. The first public demonstration of amateur television on a closed circuit took place on 21 April 1950 at an open meeting of Shefford & District Short Wave Society. The proceedings of the meeting were televised using Ivan Howard G2DUS's 250-line Iconoscope camera equipment and then shown to the 250 strong audience on a 15in tube placed in front of a screen. They filmed an usual meeting that involved a short talk on the British Amateur Television Club, a sketch by a local comedian and a live junk sale which were all seen by the audience. The press reports were most enthusiastic – one suggested the results were "better than BBC quality".

When the Postmaster General permitted amateur television on 70cm, Jeremy and Ralph Royle G2WJ/T were soon active, Jeremy designed and constructed the necessary television equipment and the transmitter and aerial was made by his father, Ralph. They refined their projects until, in July 1953, they carried out tests using a transmitter based on the CV53 valve. These first pictures were received over a distance of 31 miles by L V Dent G3GDR.

Centenary

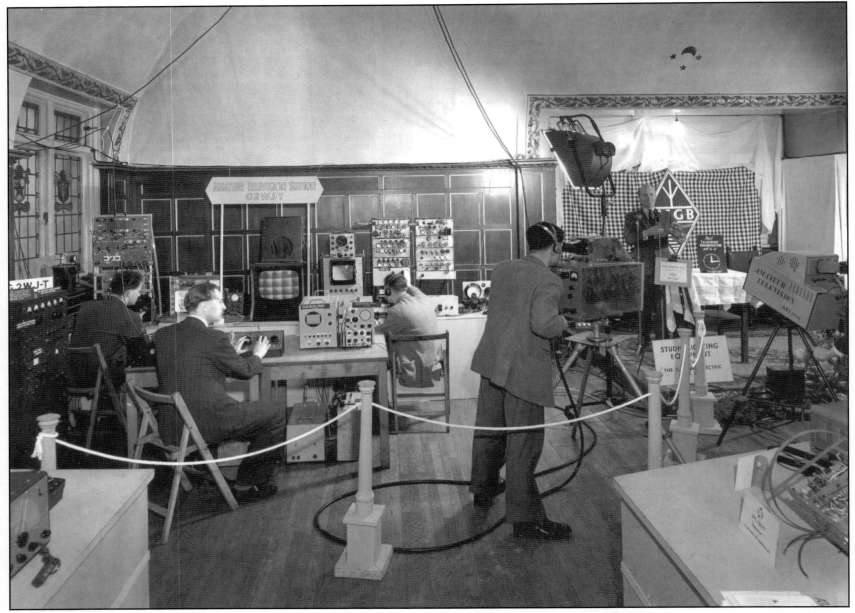

ATV stand at the RSGB Amateur Radio Exhibition, held at the Royal Hotel in London.

3. Inter War Years

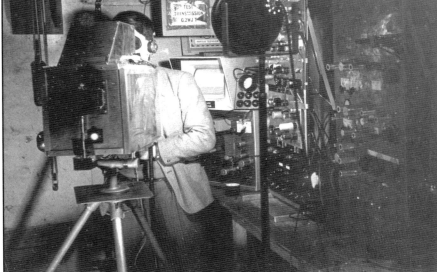

G2WJ/T television equipment.

Work immediately started on a high power transmitter and modulator. By February the following year, the new 20W transmitter was ready. Using 145.66MHz for the sound channel, pictures were reliably transmitted over the 31 mile distance. Transmissions of pictures and sound were regularly broadcast every Saturday evening between 1800 and 1900GMT. Amateur television was demonstrated on 436MHz by Ralph and Jeremy Royle G2WJ/T at the 8th Annual Amateur Radio Exhibition in November 1954. Then, in 1956, Ralph and Jeremy Royle, using the callsign G2WJ/T at Hadens End, Essex, and C Grant Dixon operating from the home of M W S Barlow G3CVO/T in Great Baddow, transmitted colour television pictures 13 miles.

Monochrome 200 line seg. Test card title.

John Clarricoats G6CL

John Clarricoats G6CL became Secretary of the RSGB in 1932, a post he held until his retirement in 1963. On 1 January 1937 he also became responsible for the general editing of the *T & R Bulletin*.

His interest in telecommunications began in 1912 when he joined the Telephone Inspection Department of the Western Electric Company. Then in 1917 he joined the Royal Flying Corps as a Wireless Operator seeing active service in both Belgium and Northern France for two years. He returned to Western Electric at the end of WWI. Clarricoats obtained his callsign in 1926 having previously operated the Company's station 2YZ. His first station ran just 7W from dry batteries but he still managed to work over 3000 other amateurs, several hundred of whom were in the USA. By 1933 he had improved his station and was running 80W and was active on 14 and 28MHz daily.

He became a member of the RSGB in 1926, he also joined the T & R Section. From 1932 until the war years, it is doubtful that he had more than half a dozen weekends at home a year. He attended and spoke at countless meetings all over the country; he could always be relied upon to turn up at a couple of dozen NFD stations and, even when at home, that became open house to radio amateurs from far and wide.

He worked tirelessly for the Society during the war years, keeping the *T & R Bulletin* running and writing several books. He held a special commission in the RAF and was on special duties connected with the training of RAF personnel in radio communications. Following WWII he was involved in many international conferences and was a member of the Postmaster General's Frequency Advisory Committee.

John Clarricoats G6CL. Station from September 1928.

A E (Arthur) Watts G6UN

Arthur Watts served, with distinction, in the Royal Navy during WWI, had lost a leg at Gallipoli and had worked during the later stages of that war on special duties at the Admiralty. He was a director of a very old established firm of cardboard box makers. In 1925 he was BRS12 later licensed as 6UN. He came to prominence in the Society with his design for a Membership certificate that was used for over 30 years. He was co-opted on to the Council in 1929 and was involved in the British Empire Radio Union. He attended the 1932 Madrid International Telecommunications Conference as an IARU observer. In 1934 he became President and served in this office for three years; in 1938 he was again elected President for a further three years. During this second period he was involved in the Society's part of the formation of the Royal Air Force's Civilian Wireless Reserve (CWR). In 1938 he attended the Cairo International Telecommunications Conference on the Society's behalf. Arthur Watts played a prominent role in organising the Voluntary Interceptors in WWII.

Centenary

RSGB Council 1934. Back Row (l-r) 2NH, 5VM, 2CX, 5YK, 6LL, 6CL. Front Row (l-r) 2TI, 5AR, 6UN, 6NF, 6UT.

IDEAS FOR EMERGENCY COMMUNICATIONS

A small article in *the Times* in April 1937 stirred up the amateur community. "A scheme for the establishment of a chain of amateur radio transmitters for use in emergency has been submitted to the Home Office, and it is expected that tests will soon be made. The plan is the outcome of experience gained during the recent floods in the Fen district, when amateurs were able to provide means of communications between isolated parts. Experimenters in the north of England worked on a scheme, and sought cooperation of other amateurs in different parts of the country, the response being so encouraging that the scheme was approved for submission to the Home Office." Although the plan didn't come directly from the RSGB (it probably came from Austin Forsyth G6PO, Editor of *Short Wave Magazine*) it did receive an enthusiastic response from RSGB Members. Sadly, with the restrictions that were brought in at the outbreak of WWII it was some 16 years before anything like this scheme came to pass.

RSGB Convention late 1930s. Taken outside Savoy Place (IEE) front row centre John Clarricoats G6CL, Arthur Watts G6UN, Ernest Dawson Ostermeyer G5AR (President).

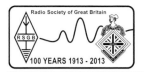

4

World War II

AMATEUR RADIO HANDBOOK

November 1938 saw publication of the first edition of the *Amateur Radio Handbook*. Despite concerns by the Council and Technical Committee that it wouldn't be popular, the first printing of 5,000 copies sold out. It had 240 pages and hundreds of illustrations, the biggest project the Society had attempted. The 24 chapters contained both technical and topical information and was the first British publication of its kind.

A second printing of 3,000 copies arrived the day before WWII was declared, fortunately these were snapped up by the Services as a valuable text book. It seemed at one point that paper rationing would stop production but the Paper Controller gave the all clear for paper to be made available after he was shown the value of the publication to all three Services.

In total, over 181,000 copies of this handbook were printed between 1938 and 1946.

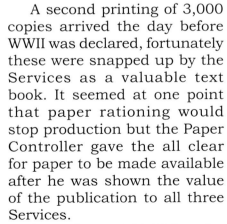

OUTBREAK OF WAR

A notice in the *London Gazette* on 31 August 1939 – and broadcast on the BBC nine o'clock news – stated that "all licences for the establishment of wireless telegraph sending and receiving for experimental purposes are hereby withdrawn". The RSGB continued its activities despite the war.

Many of the wartime activities of RSGB Members still remain unrecorded. Very few amateurs did not make use of their technical or

The *London Gazette* notice withdrawing amateur licences.

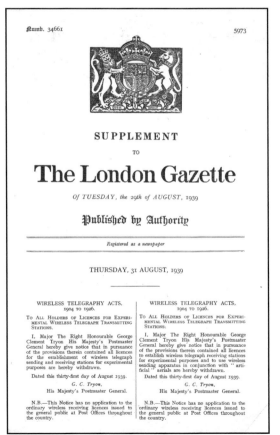

A receipt from the GPO for A W Knight's radio equipment handed in at the start of WWII. A W Knight was G2LP.

operating skills in some way or another in service of their country.

Some amateurs had joined the Royal Naval Volunteer Wireless Reserve early in the 1930s. An RAF Civilian Wireless Reserve was formed in 1938 with the cooperation of the RSGB. Civilian Wireless Reserve members, whose skills were primarily technical rather than telegraphy, were formed into RAF Emergency Fitting parties installing radio and radar equipment.

It was decided to continue to publish the *T & R Bulletin* (or the *RSGB Bulletin* as it became in 1942) throughout the war, despite shortages of paper. One of the main reasons was to provide a medium that would enable Members to keep in touch and to provide valuable reading material to those in the services. Khaki & Blue was a new feature in the *T & R Bulletin* where they published information concerning RSGB Members serving in HM Forces. The feature recorded promotions, messages and even good news such as weddings and the recovery of injured Members. Sadly it also included obituaries.

The first known war casualties among RSGB Members were Jack Hamilton G5JH and Kenneth Abbott G3JY, both lost their lives when *HMS Courageous* struck a mine on 15 September 1939. They had been drafted to *HMS Courageous* as telegraphists. On 15 September 1939, a convoy contact was made due west of the English Channel, in an area the British called the Western Approaches. The sea lanes were abuzz with traffic and some successes against British shipping had occurred in the early days of the war. To provide at least some form of protection for these ships, the Admiralty had deployed the old aircraft carrier HMS Courageous with a destroyer escort screen to conduct anti-submarine patrols. The *Courageous* sank in less than 15 minutes with the loss of 519 lives, including her commander Captain W T Makeig-Jones. Her total complement was 1,260 officers and ratings (including air group), and two squadrons of Fairey Swordfish

L E Newnham G6NZ

As Squadron Leader, Newnham was stationed at No 1 Radio School, Cranwell. In March 1940 he organised a wartime meeting of amateur radio RAF personnel and brought together service men stationed at Cranwell and the surrounding area. As President in 1958, he attended the dinner in October 1958 held for those amateurs who had held a transmitting licence for at least 25 years; this was the start of the Radio Amateurs Old Timers Association. He also attended the Geneva Radio Conference in 1959 as an advisor on amateur radio matters.

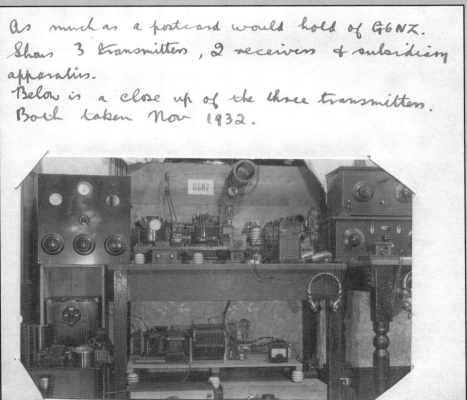

Len Newnham G6NZ and the station he operated just before WWII.
It shows three transmitters, three receivers and ancillary equipment.

aircraft (48 planes). The Veendam and a British freighter Collingsworth participated in the rescue, fishing survivors from the oily waters. The obituaries of G5JH and G3JY appeared in the October and November issues of the *T & R Bulletin*.

Some excerpts from the Khaki & Blue columns are shown below. "Congratulations to Sq Ldr Viscount Carlow G6XX and to Lady Carlow on the birth of a second son. Lord Carlow is serving as an Air Attaché to a European Legation". Sadly, Khaki & Blue also reported Carlow's death, aged just 36, in April 1944. He had been interested in Society business until the time of his death. His knowledge of radio brought him recognition in the early days of the RAF Auxiliary Reserve. For some years he was Signals Officer to No 600 Squadron based at Hendon. After the Battle of Britain he was appointed Air Attaché to the British Embassy in Helsinki and was promoted to Wing Commander. Later he acted in a similar capacity to the British Embassy in Rio de Janeiro where he was promoted to Acting Air Commodore. He visited many amateur stations throughout Europe and was an exceptional ambassador for the Society.

"We understand that Arthur Simmonds G3AD has been promoted to rank of LAC"

"R G Shears G8KW, who is serving as a signal man in the R C of S wishes to be remembered to old friends"

"As an example of real ham spirit, 5 or 6 weeks ago I received a letter from G8--of Suffolk forwarded from the RSGB. I wrote to him ... since then he has written to me nearly every week and has sent me parcels of books, magazines, a French Grammar, mouth organ, petrol lighter and other useful things from a chap I have only met once"

"The many friends of Lt Bill Brigden, RNVR, (G6WU) will be glad to hear he is making a good recovery after being wounded recently in a North Sea engagement"

"When I first went to France in January, I took with me a few old "Bulls". Thank goodness I did as they were absolutely invaluable to me in Dunkirk. I was waiting to get off for nearly four days and I read them from cover to cover during that period. Conditions were slightly trying and the "Bull" took me into the past for a brief while. Unfortunately, these valued copies had to remain on the beach with the rest of my kit". You'll be pleased to know that the RSGB sent L Frank G4NU a new set of magazines.

"The many friends of Capt R H B Candow RAOC (GM3SC) will be pleased to hear that although in the water for three hours after the Lancastria went down, he arrived back safely to England little worse for his experience"

"Congratulations to F/Lt Austin Forsyth G6FO of Newport, Mon on his recent marriage to ASO M E Coates WAAF of Sutton Coldfield" Austin Forsyth G6FO was editor of *Short Wave Magazine* for many years.

THE EARLY BIRDS

By 1938, British Armed Services in Britain realised that, in the event of war, there could well be a shortage of wireless operators. With the help of the RSGB, the Civilian Wireless Reserve (CWR) was created and radio amateurs were encouraged to join. Later, the CWR was absorbed into the RAF Volunteer Reserve. They were one of the first groups to set foot on the Continent in WWII. Nicknamed The Early Birds in the *T & R Bulletin*, of the 52 personnel more than 40 were licensed amateurs. They were under the command of F/Lt C S Goode G2OH and Sgt S Leslie Hill G8KS was a senior NCO. Messages home were often reported in the Khaki & Blue column.

Each station had two operators and the gear was the T1083 / R1082 duo, powered by lead-acid batteries and a small petrol-electric set for charging. The antenna system consisted of a random length inverted L, described by those that used them as "the AOG (Act of God - because no-one knew how the hell it worked)". The antenna was supported by two guyed steel masts that came in four foot sections in a canvas bag with assorted guy ropes.

4. World War II

Reunion dinner of the Early Birds.

The RAF Section of the first Wireless Intelligence Screen unit on parade in the Caserne Ney at Metz in north eastern France during the spring of 1940.

Victor Desmond G5VM

Victor Desmond G5VM was placed 4th amongst UK stations in the first Empire Contest in 1931. He was on the committee set up in December 1935 to prepare for the Cairo Conference representing amateur radio interests. G5VM organised the first wartime Provincial Meeting held in April 1940 in Birmingham when over 100 members including many stationed at Army and RAF establishments attended. He was the first Provincial member to be elected President in 1948 - 1949. First licensed in 1928, he was a volunteer for the Society from 1930 onwards. Prior to the war he attended almost every official RSGB meeting within a radius of about 150 miles of Birmingham using his own light aircraft for travelling from place to place. During his Presidential address he donated two trophies for work on the proposed 420-460MHz band. He was one of the first amateurs to appreciate the advantages of SSB and his station used that mode from an early date. He was elected an Honorary Member in 1952.

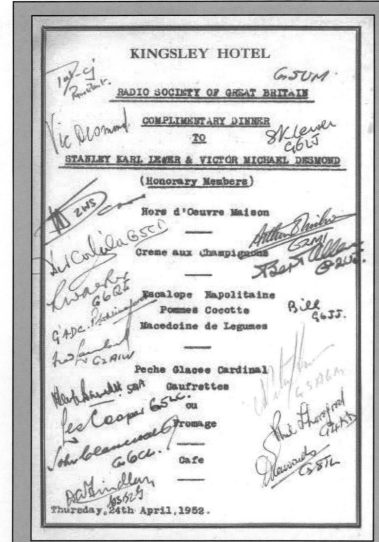

Autographed menu card from the dinner held to celebrate G5VM being elected an Honorary Member in 1952.

VOLUNTARY INTERCEPTORS

In 1939, Lord Sandhurst of MI5, approached the President of the RSGB, Arthur Watts, to determine whether radio amateurs could provide a listening watch for the Intelligence Services. Radio amateurs were often good at reading weak Morse signals and therefore many were recruited as Voluntary Interceptors (VI).

These radio enthusiasts who were, for one reason and another, unable to serve in the armed forces, were asked if they would undertake some voluntary 'work of national importance'. They were given no clue what this 'work' would be until they had signed the official Secrets Act and become Voluntary Interceptors working for the Radio Security Service (RSS). After signing they were briefed to listen for and report any unusual transmissions – all carried out in secret. Initially they were to listen for enemy agents transmitting within the UK, later they listened on HF for anything they could not recognise as genuine commercial or military transmissions. They wrote down the Morse signals received on blank log sheets provided by the Radio Security Service and posted these reports to Box 25, Barnet, Herts. It was here that the intercepted traffic was sorted and analysed and the volunteers' efforts were coordinated.

Amongst the signals heard there were some strange stations using amateur type signals but with 3 letter callsigns and sending messages in 5 letter groups of code, very unamateur like. These signals were worked on and it was discovered that they were in a hand cipher that translated into German. The translation showed that they were not ordinary signals but apparently coming from the German Secret Service. Encoded messages were sent to Bletchley Park for decoding and onward transmission to the Allied Commanders and the Prime Minister, Winston Churchill. There was a lot to be uncovered about these secret transmissions and amateurs were recruited as VIs on an ever-increasing scale – around 1,700 were eventually engaged on intercepting, although not

A Voluntary Interceptors meeting held at the home of Nell Corry on 26 August 1945

A Voluntary Interceptors meeting held at the home of H Johnson G2XC on 29 September 1945.

all at one time as many were called up for other duties. Their work has only recently been recognised.

OPERATION FLYPAPER

The desire of many amateurs to get back on the air was partly satisfied, for eight at least, who took part in two mysterious operations in 1944 and 1945. The callsigns G7FA to G7FJ were allocated to some well known radio amateurs including Morton Evans G5KJ, R L Varney G5RV, S Riesen G5SR, K G R Lee G6GL and R W Addie G8LT for Operation Flypaper. As the war was coming to a close it was felt they could glean some interesting information from foreign amateurs. They were allowed to operate from 3.5 – 4, 7 – 7.3, 14 – 14.4 and 28 – 30MHz and power was not to exceed 50W. The rules under which they could operate were comprehensive with details on what could be said and what information could be exchanged outlined in a memo they all received. For example, in the event of anyone asking for QRA, only the county could be given as a reply and also the subject of the weather was completely banned. "If one of the group is asked about the weather he must reply 'not allowed'". It's not obvious whether anything useful came of this experiment.

The Wilton Scheme took place for 3 months in 1945. The aim was to make contact with prisoners of war (POW) to gain information about their circumstances. POWs were known to have constructed receivers and, in some extreme cases, transmitters. It seems that no contact was made by the G7s who took part.

INTERNATIONAL FRIENDSHIP

Amateur radio has always meant international friendship. Never was that more clearly demonstrated than during the WWII. In February 1941, the Managing Secretary of the ARRL, Ken Warner W1EH, wrote to the RSGB concerned over the 'difficulties in the food situation' or rationing to those in Britain. ARRL President, George Bailey W1KH, and ARRL staff decided to do something to help their fellow amateurs and promptly sent cocoa, marmalade, granulated sugar and several pounds of tea across the Atlantic. They sent it in two parcels, a few days apart, valued at just $5 and $6 each. Warner was most concerned about the chances of the packages arriving safely and encouraged Clarricoats to write back and let them know so they could send more parcels. He also reported that it looked like most of September and December *QST* magazines destined for the UK had been lost at sea. Considering the censorship on letters leaving the UK, it is surprising to read very political comments and his views on whether America would enter the war.

In April, Clarricoats wrote back saying that both parcels had arrived safely and how much they were appreciated.

Warner stated in his May letter that, "With that successful experience behind us, we want to try it some more". So the next package was prepared and this contained tinned meat (three tins of chicken, two tins of ox tongue, one of roast beef and one of frankfurters), almost $9 worth. The letter also talked about how national defence matters had started appearing in QST, one called 'In the Services' as nearly 3,000 amateurs were Naval reservists, with a similar number in the Army. The other section was called 'USA Calling' and was a compilation of the calls of the armed forces and civilian agencies that need radio personnel likely to come from amateur ranks. One interesting item was the problems with interference from two BBC stations in the 7MHz band. The stations were putting out Empire News and the BBC had no plans to stop the transmissions. Warner noted that, "If BBC is using directive antennas, it is obvious that something is out of kilter and they are losing a lot of power behind the

RSGB wartime Convention held on 9 August 1941.
The front row contains Nell Corry G2YL, Barbara Dunn G6YL, John Clarricoats G6CL and Gerry Marcuse G2NM.

antenna that is not doing useful work for them.... so I am endeavouring to find out whether it is in order to make a technical report on the matter which might result in readjustment of the transmitter in such a manner as to cut down the American interference".

A couple of months later, Clarricoats wrote back saying the parcel of tinned meat had arrived safely and that if the ARRL hadn't sent earlier parcels of tea and marmalade, they would have been well out of stock. He also mentioned that the cocoa that had been sent "was much appreciated because it was streets ahead of anything we get here". It's interesting to read about the effects of censorship on the *T & R Bulletin* and how he felt that it will make things stilted.

Sadly, not all the parcels got through – and it wasn't because of the U-boats either. Some parcels were seized by H M Customs & Excise because they contained more than 2lb of tea, the maximum you were allowed. Undeterred, the ARRL promptly sent more parcels with 1½lb of tea plus 6 large chocolate bars and a small bag of candy. Clarricoats was also pleased to discover that the BBC had removed the station broadcasting on 7150kHz and so the US stations had much better reception. Apparently, the BBC had been troubled with an echo from the antenna that the ARRL had identified as the culprit.

Help in the way of food parcels and food vouchers didn't only come from the USA. After WWII ended, the Wireless Institute of Australia (WIA), Queensland Division sent parcels to the UK and a ballot was held to choose the Members that received these parcels. Many letters of thanks were sent back to Australia with every delivery. The National Society of South Africa joined WIA and the ARRL in sending these very welcome parcels until the 1950s. Another club helping UK amateurs during those difficult days was the Ontario Phone Club. Many Members were very thankful for the generosity of other amateurs around the world. These acts were truly gestures of real amateur friendship.

POW FUND

The RSGB didn't only receive charity from overseas, they played their part too. In 1940, the Society started the RSGB Prisoners of War Fund to which Members contributed varying amounts. In the five years it was running they sent 1,329 parcels to prisoners of war in Europe. That meant 357,430 cigarettes, 8,480 ounces of tobacco, 2,007 new books, 99 games and 23 sets of music. When the fund closed to new donations, £700 was left in the account for parcels to POWs in Japan as they hadn't been able to send parcels to the Far East previously. Many letters were received thanking fellow Members for their generosity. "I wish to express my gratitude to you – and through you to the Members of the RSGB – for the incredible generosity of the Society in sending me parcels during my four years as a prisoner of war."

"I have received from the Society thousands of cigarettes, many books, games, a most welcome kit bag and case. Such generosity is quite overwhelming and it is impossible for me, in words suitable, to express my gratitude."

"The cigarettes were not only sufficient for my own needs but in times of acute shortage, I was able to help my friends who were as amazed as I was myself at the open handedness of the Fairy Godmother of a Society. The books I read and passed on to others to enjoy and the kit bag was very timely in its arrival."

"These gifts from the RSGB have materially lightened the burden of prisoner of war existence, not only for myself, but for those about me. I have always made it clear that these parcels I constantly received were gifts from the RSGB."

"It is wonderful to be home again and I am glad to say I am 100% fit." D Gordon Blair G8VU.

Amateur radio meeting held at the Tadworth, Surrey, home of Nell Corry G2YL in August 1940.
She had Lord Sandhurst from the RSS as one of her guests. If you look closely you can see a copy of the Amateur Radio Handbook right at the front.

Centenary

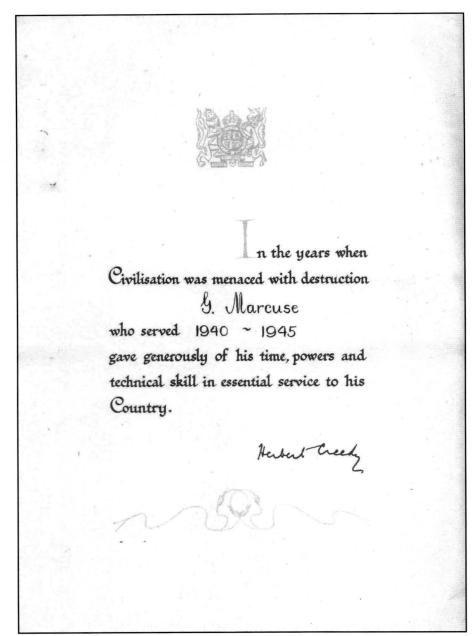

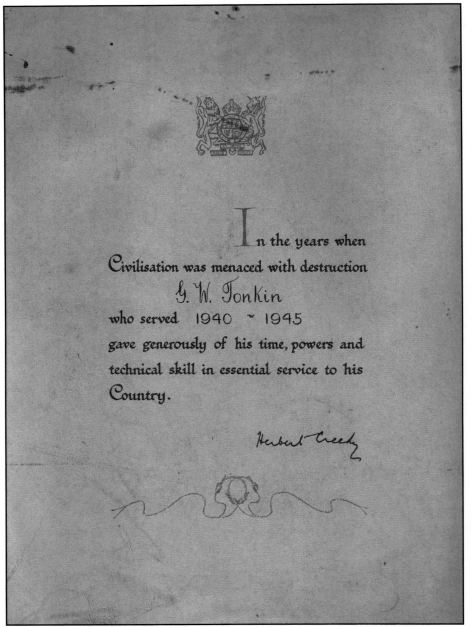

Post War to Century's End

BACK ON THE AIR

The GPO stated its intention to restore facilities to all pre-war fully licensed amateurs well before war had ended. They said that pre-war callsigns would be reallocated – unless a different callsign was requested. The GPO also agreed the RSGB's proposal that those who had served in the radio trade in the Services, with proof of proficiency, would be granted a licence. All new licensees would be required to pass a Morse test and a technical exam. Those passing would receive a callsign with the prefix G3 and 3 letters, such as G3ABC. The RSGB was closely associated with the work on these new exams and the syllabus. The first exam was held on 8 May 1946 at a number of centres across the country. A second exam was held in November of that year.

Following the end of WWII, in November 1945, the GPO released the bands 28.000 and 29.000MHz and 58.500-60MHz to radio amateurs who had renewed their pre-war licences. They were permitted to use 100W on 28MHz and 25W on the 58.5MHz band. Those who had upgraded their pre-war non-radiating licences could only use 25W, subject to passing a Morse test. The 7.150-7.300MHz and 14.100-14.300MHz bands were released in June 1946. Gradually more bands were added as things returned to normal.

C G (Bert) Allen G8IG

C G Allen G8IG of Bromley, Kent made one of the first authorised post-war amateur contacts between England and the Continent when he worked the Norwegian station LA8C during the latter part of January 1946 on 28MHz. Bert was also one of the first six holders of the Empire DX Certificate, awarded to those who could submit evidence of having established two-way contacts on 14MHz with amateur stations situated on 50 different Dominion or Colonial call areas and also two way contacts with amateur stations in 50 different Dominion or Colonial call areas on bands other than 14MHz.

F C (Dud) Charman BEM G6CJ

In the mid-1920s, many authorities were sceptical of the existence of a reflective layer in the upper atmosphere that had been postulated by Kennelly and Heaviside. However, an experiment by Appleton and Barnet in 1925 using the BBC's long wave transmitter gave evidence of reflection. In 1927, there was a total eclipse of the sun, with the shadow passing across Northern England. It occurred to Charman and H Clark G6OT – who were both working in Bedford at the time – that this would be a good opportunity to attempt to measure the height of the reflective layer. They arranged with G5YG of Glasgow to radiate a carrier on 160m for several days, including the day of the eclipse. A suitable receiver was used, having two stages of neutralised HF amplifiers, giving enough output to operate a milliammeter. A record was made using a roll of a paper tape over a frame with a transverse narrow slot, so that the observer could follow the meter pointer; the timing at one minute intervals was arranged to mark the paper by a foot pedal. The signal from Glasgow was received on the days prior to the eclipse and was fairly consistent. When the eclipse took place, very good interference patterns were obtained at the beginning and end of the event. Charman and Clark calculated that the height of the layer was 50 – 100 miles.

G6CJ wrote a stream of articles in the *T & R Bulletin* as well as in the *Amateur Radio Handbook*. In early 1941, on temporary secondment from EMI to the Radio Security Service, he developed the pioneering, wideband aerial distribution amplifiers that were installed at Hanslope Park and many of the special intercept stations of the Radio Security Service (RSS) and MI6's Special Communications Unit (SCU). He was one of a small group of Voluntary Interceptors, under the cover of the Royal Observer Corps, who received the British Empire Medal in 1946.

In most branches of radio it is possible to bring some kind of practical demonstration into the lecture hall. In the case of aerials, the speaker – until G6CJ and his Aerial Circus – had been forced to confine his remarks to theoretical considerations. Dud G6CJ designed a system that used microwaves to make a demonstration possible. He first demonstrated the equipment at a meeting of the Society in 1945. Using micro-mini aerial models he was able to illustrate exactly what happens to the RF when, for example, reflectors and directors are added to a simple dipole.

He was President of the Society in 1952.

Dud Charman G6CJ operating VE1WL during the 1966 BERU Contest

"Somewhere in Britain in 1942" says the photo album. Dud designed wideband aerial distribution amplifiers for the Y stations. This shows the lab benches at Hanslope Park

Dud, seen here with Gerry Marcuse G2NM, Irene Marcuse and G Exeter G6YK.

Dud and his Aerial Circus.

Aerial views of Hanslope Park in the post-war period.

Centenary

G6YB/P during NFD 1995, the Bristol Contest Group.

A herd of young bullocks added country atmosphere to the NFD scene at Verulam's G2AIA/P station.

Wellies and a bikini – it can only mean NFD! G4DJX and G4JKS operating the Verulam ARC entry.

NFD 1968. R Channahan (GPO Radio Services Division) putting the Ayrshire ARG B station under the inspection microscope in the early hours of Sunday morning accompanied by GM3UVK.

MORE NATIONAL FIELD DAYS

The first post war National Field Day was held in June 1947. Nearly 200 individual portable stations were involved with something like 1,500 operators and helpers taking part. Once again, the weather was reported as 'squally'! Southgate ARS won with their G5FA/P and G6ZO/P stations. G5FA was running the 3.5 and 1.7MHz bands stations with both feeding a 264ft inverted L aerial running in a north-east to south-west direction. They finished with a total score of 244. G6ZO ran the 7 and 14MHz stations and a 400ft long wire supplementing the 7 and 14MHz dipoles. This station scored 339 points. Joint second placed was Coventry with G2YS and G5PP stations and Cambridge with G5JO and G5DQ. Several European countries operated portable stations in the field; the Swiss showed the keenest interest. HB1JJ submitted a log having worked all 114 G portables

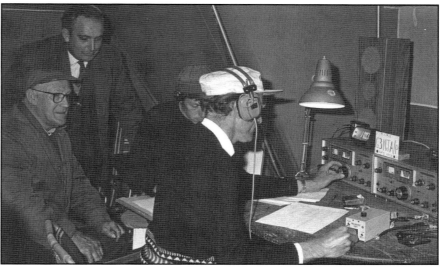

Torbay NFD 1972.

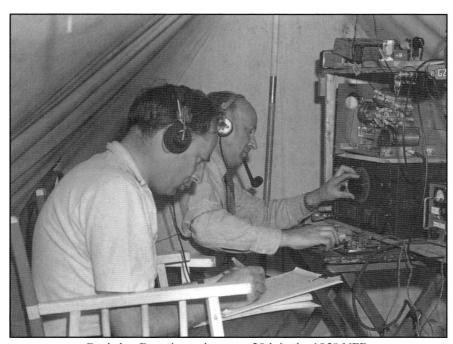

Coulsdon D station, who came 29th in the 1958 NFD.

Margaret Mills G3ACC at the Dulwich & New Cross B station at Crystal Palace.

50MHz PERMITS

A limited number of 50-54MHz band permits were issued by the GPO in November 1947 and the work conducted by these experimenters really paid off. Following this, a period of crossband contacts (50/28MHz) took place across the Atlantic. Denis Heightman G6DH had decided the only way to improve the chance of transatlantic QSOs when the permission was granted would be to build a beam antenna. He completed his 3-element beam from scratch and mounted it some 38ft above the ground. On 5 November, W1HDQ called on 50MHz and worked G6LK crossband to 28MHz. Denis G6DH then gave him a quick call and, as he had one of the new 50MHz permits, asked the American to "listen up on 50MHz right away". At 1301GMT, Denis called W1HDQ and he replied immediately with an S7 report – the first 50MHz transatlantic contact from England. The contact was very good but they kept it to about 10 minutes because others were waiting to make the transatlantic contact. Denis went on to work W2AMJ, W1LL, W8MVG, W1AF and W1CLS.

A few days later, Denis called Major Ken Ellis G5KW who was operating from the Suez Canal under the callsign MD5KW. They managed a quick telegraphy contact before changing to 'phone and reporting a "fantastic 20dB over 9" contact.

Throughout the 50MHz record breaking contacts, his wife, Eileen, could be relied on to provide meals, often at very odd times, according to the prevailing propagation conditions.

Forces Broadcasting Service HQ team Maintenance Unit. Ken Ellis G5KW (centre of the back row) was heard operating from the Suez Canal under the callsign MD5KW.

5. Post War to Century's End

Robert G D Holmes G6RH

Robert G6RH was one of the first six holders of the Empire DX Certificate that was hand-produced on vellum. This was awarded to those who could submit evidence of having established two-way contacts on 14MHz with amateur stations in 50 different Dominion or Colonial Call Areas and also two-way contacts in 50 different Dominion or Colonial call areas on bands other than 14MHz. More than 100 Dominion and Colonial call areas were in existence circa 1946.

GB1RS ON THE AIR

An RSGB Headquarters' station made its début at 2000BST on 1 September 1948. GB1RS made regular transmissions of standard frequency signals each hour from 0600 to 2400GMT. The message, in CW, was CQ de GB1RS (repeated) QRG 3500.25kc/s. VA GB1RS (followed by a long dash).

The frequency was chosen as it provided a marker that, together with its harmonics, gave points below which an amateur transmitter working on the 3.5, 7, 14, 21 and 28MHz bands couldn't be considered 'safely inside'. The transmitter was donated by EMI Ltd and had an output of 350W from two 813 valves. The aerial used two heavy duty masts at the ends of the building that could be used on 7MHz or as a T aerial against the earth from the metal work on the roof on 3.5MHz. Later the times were changed to 1800 to 0800GMT.

GB1RS, the HQ station.

Stanley Karl Lewer BSc G6LJ

Stanley Karl Lewer BSc G6LJ represented the RSGB at the 1st IARU meeting and sat on the Callsigns' Committee in 1925. The day after he was made RSGB President, in 1947, a dinner was held in London, given by the Council to Society Members who had been prisoners of war. He attended the World Radio Conference in Atlantic City in 1947 – he was there for 4 months and largely as a result of efforts made by RSGB/IARU observers, European amateurs emerged from the Conference little worse for wear except around 7MHz, which was reduced by half. They did manage to retain the 1.7-2MHz band. He was made an Honorary Member in 1951.

The first post-war Amateur Radio Exhibition in the Royal Hotel, London was held in November 1947.

EXHIBITIONS

The first post-war Amateur Radio Exhibition, in 1949, was opened by Col Sir Stanley Angwin, Chairman of Cable and Wireless Ltd. Around 5,000 people visited the show and many Members travelled long distances to attend and see the wide range of British equipment on show. The RSGB stand was an exhibition of equipment loaned by members of the Technical Committee. It included a valve voltmeter with diode probe by G2IG, a reactance modulator for narrow band FM by G5CD, a 25W transmitter by G6LL, a 150W 6J5-807-808 transmitter by G2MI and an amateur band receiver by G8PD amongst the exhibits.

FREQUENCY LOSSES & ADDITIONS

The Postmaster General withdrew the 58.8-60MHz band at midnight on 31 March 1949. But it went with flying colours! At the end of the 31st, the band apparently 'sounded like a well-supported contest'. Users of the band applied their experience of the 58MHz band to the 144MHz band instead.

The Postmaster General made a surprise announcement in the House of Commons on 25 October 1950. He decided to licence transmissions by amateur television signals in the 2.3, 5.65 and 10GHz bands. Whilst, at the time, amateurs didn't feel these were the most suitable for the job, they were once again pleased to have the opportunity to transmit TV.

Amateurs were given permission to operate /P within a 10 mile distance from a named point in May 1950. The permits were issued subject to the normal licence conditions, although power was restricted to 25W (10W on the 1.8MHz band) and with the payment of 10 shillings. This permit did not apply to official NFD permits that were issued free of charge.

IARU MEETINGS

In May 1950, a delegation from the Society took part in the 25th Anniversary Congress of the IARU. Around 100 delegates from 15 IARU Member Societies assembled in Geneva for the Congress. The meeting agreed that band planning as a principle should be accepted and, while some delegates spoke in favour of compulsory band planning, it was generally agreed that it should take the form of a recommendation. It was also agreed that the number of international contests should be reduced by combining certain Member Society's contests such as Field Days.

From this, the triennial IARU conferences were to emerge and the RSGB takes a full role in proceedings today. In 1960 and 1981, the Society hosted two of these triennial meetings. Today, the International Amateur Radio Union is composed of 162 national member societies.

IARU 25th Anniversary Congress.

William A Scarr G2WS

In the mid-1930s, William Scarr G2WS was well-known for walking the hills of Derbyshire carrying a compact 5m station. Later, the first recorded instance of 420MHz portable operation within the UK took place on 22 January 1949 when William Scarr G2WS operating from the North Surrey Downs made two-way contact with Charles Newton G2FKZ (Dulwich) 12 miles away. On 19 June 1949, G2WS, G2FKZ and others helped to establish new records on this band when distances up to 40 miles were worked. A month later, G2WS/P operating portable from Charing, near Ashford, contacted G3BEX/P on the Devils Dyke near Brighton, 50 miles away.

The G2WS home station in 1931.

William Scarr G2WS with his compact 5m portable station in the mid 1930s

G2WS with his 70cm station during an RSGB contest in August 1949, located in Ashdown Forest

CALLBOOK

The first edition of the *RSGB Amateur Radio Callbook* (now called the *RSGB Yearbook*) was printed in 1951. It listed over 6,000 callsigns, names and addresses of amateur transmitting stations in the British Isles and Eire. Priced at 3/6, it was edited by J Tyndall G2QI.

WHEN ALL ELSE FAILS THERE'S AMATEUR RADIO

In the final hours of January 1953, devastating flooding struck the coasts of the British Isles. Telephones, government wireless stations and the utility services were put out of action for days. Radio amateurs in those stricken areas immediately placed their stations at the disposal of the authorities – even though, strictly speaking, this was against the terms of the licence.

In Lincolnshire, for example, when Humber Radio coast station was put out of action by the floods, local amateurs maintained a continuous watch on the shipping frequencies. Four times in a few hours, Grimsby and Hull Radio Amateurs intercepted distress calls from ships at sea. All this was after the GPO had turned down the idea of an amateur emergency network!

Reg Hutcheson-Collins G3AXS was maintaining a silent watch on the shipping frequency and heard Humber Radio in contact with the *SS Levenwood*. The ship required the aid of tugs and urgent medical advice for the First Mate. Humber Radio acknowledged the call and asked North Foreland to listen on 164m. On 1825kHz, Humber asked North Foreland to deal with the situation as their landlines were down – then silence. Humber Radio had closed down involuntarily. *SS Levenwood* was still calling and getting no reply. So, after some hesitation, G3AXS tuned to 1.650MHz and called *Levenwood* and asked if he could help. He then contacted the local hospital and got the necessary medical

RAYNET stand at the Scottish Convention in 1990. **Photo by GM4SRL**

advice and passed it on to the Master of the *Levenwood*. Then G3AXS arranged for a tug to go to the assistance of the ship. Eventually, the *Levenwood* reported that the vessel was under control but would he maintain a watch. At around midnight, the Master reported that the First Officer was now comfortable but still requested the watch to be kept longer. The *MV Menapia* broke adrift from the tug assisting her and couldn't establish contact with the tug. G3AXS was able to telephone the owners of the tug and reported the situation to Cullercoates Radio. During the next few hours, G3AXS helped the *Humber Light Vessel* and the SS *Melrose Abbey*. Four distress calls aided by the quick thinking of an amateur radio station that was on the air when the normal communications systems went down.

The 144MHz Field Day on 1962 was remembered for more than the competition. For many, the weather conditions were "very adverse with a howling gale". Northampton Short Wave Club were on the Isle of Wight with a steadily mounting score when, at 4pm, a man staggered into the station. With bleeding hands and torn clothes, he told the amateurs that a DC3 aircraft had crashed into the hillside about 200 yards away. G2HCG put a call out using G3GWM/P, which was heard by G3NIM of Southampton. He called the emergency services and within five minutes of the call, sirens could be heard. Meanwhile, the rest of the team went to the crash site to assist the injured where they gave up their coats to make the survivors as comfortable as possible. The club decided to move their station as

G3VJB/M receiving instructions before moving off at Newquay during an oil emergency.

Joe G3THT at the control station in Newquay council chamber.

G2HCG/M to the scene of the disaster with G5NF acting as the other end of the circuit. The first message they handled was for a police officer without a radio who asked the police to take full disaster action. When ambulances equipped with radios arrived, the amateur station was closed down after maintaining two VHF channels for almost an hour. After the operation, senior police officials and the Rescue Coordination Service, Plymouth conveyed their thanks to all the amateurs concerned.

In April 1975, a party of boys from Simon Langton Boys Grammar School with G3LCK in charge accompanied by G8IAM, G8IZJ and G8KCA were working in the Tal-y-Bont region. G8IZJ was operating on 2m and in QSO with GW8HCZ when two pupils from the Helena Romnes Comprehensive School arrived with news that one of their party was suffering from exposure at about 1,500ft ASL. GW8HCZ telephoned the information to the police and mountain rescue who requested they keep 145MHz open. G8KA set off to the casualty and G3GZX acted as a relay for messages. Police and mountain rescue soon arrived and both expressed their appreciation of the assistance rendered and of the overall efficiency of amateur radio as a means of emergency communication.

At Easter in 1987, Brian Tutt G4ZZK answered a Mayday call on 21MHz from a sinking yacht off Ascension Island. At 1630GMT on 18 April, he was waiting to call ZS1JD on 21.326MHz when he heard a weak and broken mayday signal. Brian called the South African station to see whether

The talk-in station at the Sandown Park RSGB VHF Convention in May 1988 was provided by members of South West London RAYNET Group.

Members of the Scottish Radio Amateurs Emergency Network.

Leslie Cooper G5LC

Leslie Cooper G5LC was RSGB President in 1953 and also President of the Thames Valley ARTS. He attended the first Garden Party at Buckingham Palace that the RSGB was invited to with Mrs Cooper, John Clarricoats G6CL and Mrs Clarricoats. G5LC travelled to Iserlohn in August 1953 to represent the Society at the annual convention of the Deutscher Amateur Radio Club (DARC) and took part in a fox hunt. He was hunting a 2m transmitter placed high up on a wooded slope about 5 miles outside Iserlohn.

At a lunch following the opening of the 7th Annual Amateur Radio Exhibition, in November 1953, he announced the formation of the Radio Amateur Emergency Network.

The station of Leslie Cooper G5LC.

he had any more information, then he rang the local Coastguard. Nothing more was heard from the sinking vessel but the next morning the Coastguard told Brian that his action had probably saved the lives of the two yachtsmen who were then safe and well on Ascension.

Richard G0OII, the keeper of the GB3YC 2m repeater, was monitoring activity on YC in May 2001 when he received a call from Simon G8PXB/M. He was working as a volunteer ranger in relation to the Foot & Mouth crisis in the nearby national park. Simon had discovered a moorland fire that was getting out of control and was assisting another ranger to beat out the flames. The fire's intensity increased and Simon attempted to alert the fire brigade by mobile phone but without success as there was no network service in that remote part of the moors. However, GB3YC was offering an 'end-stop' signal and he used the repeater to contact Richard, requesting that he call the fire brigade. This was done and six fire appliances attended the scene. Soon after, Richard was phoned by the North Yorkshire Fire & Rescue control room who said they had lost contact with their appliances as their radio system had no coverage in the area where the fire was burning. They asked if a link via the repeater could be set up with those at the scene to relay status messages so that they would know when their appliances had arrived and the state of the incident. Later the control room officer thanked the radio amateurs for their cooperation and praised the quality of service provided by GB3YC. Two hours later, Simon called back through GB3YC to reports that the fire was out and the fire brigade had left the scene.

LICENCE CHANGES

The GPO reorganised amateur licences in June 1954 with just three categories – amateur sound, sound-mobile and television licences. Operators could use their callsign at home and at any reasonably permanent alternative address without having to use /A – temporary addresses, such as holidays, still would require /A. Mobile operation was available to everyone who applied for the sound-mobile licence, which was slightly cheaper than the main licence (£1 and £2 respectively). A television licence would have cost you a further £2, which was a reduction on previous charges.

RSGB NATIONAL CONVENTION

At 10am on 17 September 1954, the 3rd post-war RSGB National Convention officially began. For the first time it was held outside of London – in Bristol. The Royal West of England Academy was taken over and turned into a huge amateur radio exhibition with stands, lecture rooms and catering facilities. It was opened by the Lord Mayor of Bristol, Alderman Kenneth Brown JP, who showed an interest in the official amateur station GB3NCB. Several trips were organised for the visitors to such places as the aircraft division of the Bristol Aeroplane Co works, British Electricity Authorities control room, BBC studios and Portishead Coast station. Lectures included TVI problems by Louis Varney G5RV and Aerials for DX by Dud Charman G6CJ. During

The GB3RS station at the 1954 RSGB National Convention in Bristol

RSGB President Arthur Mills G2MI, Alderman Brown, Herbert Bartlett G5QA and John Clarricoats G6CL inspect a copy of the Guide to Amateur Radio

Margaret Allen G3HYL, Renate Aurand DJ1YL, W5ZER and Margaret Mills G3ACC at the Bristol National Convention.

the Convention Dinner, a surprise presentation was made to the Bristol Group. The latest copy of the RSGB Bulletin was published on the eve of the Convention in which it was announced that the Bristol Group had won the 1954 NFD Contest – for the 3rd successive year!

The RSGB also received a Chain of Office, donated by Wilfred Butler G5LJ, to be worn by the President at official functions.

AMATEUR RADIO IN THE SCIENCE MUSEUM

An amateur radio demonstration station was set up in a demonstration room near the Communications Gallery at the Science Museum in London in the summer of 1955. Designed by members of the RSGB Technical Committee, it was operated by licensed members of the museum staff. Gerald Garratt G5CS was the Deputy Keeper in charge of the Communication Department and Geoff Voller G3JUL

GB2SM, the amateur radio demonstration station at the Science Museum.

was the Assistant in that department. The main purpose of the station was to demonstrate the techniques and practical operation of a radio station. By showing some of the interest and fascination of DX working, the RSGB hoped to encourage visitors to take an interest both in the hobby and consider a career in some branch of the radio industry.

The Science Museum radio station, GB2SM, celebrated its 25th anniversary in 1980. During that time, tens of thousands of contacts had been made around the world. The station progressed from a simple table-top layout into a large purpose-built console that enabled visitors to see all that was happening. During times of maximum activity, they could have three separate operating positions running so more than one mode could be demonstrated. The main position used a Collins KWM2, an 30L1 with a 75S3B and a Racal 1772 receiver also available. An alternative position used a KW2000E integrated with an Eddystone EA12. VHF operation was covered by a Trio TS700 and a linear amplifier. Geoff Voller G3JUL had remained as the staff operator since it started in 1955 but assistance had been given by a number of volunteers, all experienced operators. The station was available during museum hours and demonstrations were available Monday to Friday from 11.30am to 4pm and Sundays from 3pm to 3.50pm.

Geoff went on to celebrate his 30th anniversary as chief officer when he received congratulations from Joan Heathershaw G4CHH, the first lady President of the RSGB.

GB2RS NEWS

GB2RS first broadcast in September 1955 when Frank Hicks-Arnold G6MB read the news. The Post Office had finally agreed that the Society could broadcast short news items on Sunday mornings on 3600kHz. Later, 2m broadcasts were added. On 25 September 1955 the first RSGB News Bulletin was broadcast from the station of Council member Frank Hicks-Arnold, G6MB using the callsign GB2RS. It had taken two years of negotiations with the GPO to get to the first broadcast. Initially, the GPO had declined to allow the HQ station to be used for broadcasting purposes, despite a number of other national radio societies having a news service. Jimmy Porter, GI3GGY carried out some test transmissions from his home in Northern Ireland on 7047.5kHz using AM telephony to prove the reliability of such a service. So an experimental newsreading under his own callsign began on 14 March 1954.

At a meeting in the GPO Headquarters in July 1955, the GPO agreed to allow the Society to "broadcast news bulletins to its members". It was agreed that the broadcasts would be

On Sunday 25 September 1955, an ambition of the UK amateur radio movement was fulfilled when, at 1000GMT, the first RSGB News Bulletin was broadcast from the station of Council member Frank Hicks-Arnold G6MB, using the HQ callsign GB2RS.

Early rallies were a wonderful display of mobile amateur radio stations.

made on Sunday mornings on a frequency in the 3.5 to 3.8MHz band, and that telephony and/or telegraphy may be used. By first post the morning after the first broadcast more than 50 reception reports were received at Headquarters.

The RSGB had a series of celebratory news broadcasts to mark the 50th anniversary of its GB2RS news service. A greeting from its Patron – HRH The Prince Philip, Duke of Edinburgh – was broadcast on 25 September by all newsreaders.

Trisha Day G4KVV was the first lady GB2RS newsreader and made her début in February 1981. Her husband, G2ZYY, was the normal newsreader for listeners in East Cornwall and South Devon. Trisha and John Butcher G4GWJ were reserve readers for the area.

An early mobile rally with traders inside and outside the marquees.

Longleat Mobile Rally June 1975. **Photo by G8JDY**

RALLIES

Originally the bright idea of Brian G3GJX, the first UK mobile rally was held during October 1955 in the car park of the Perch Inn at Binsey and was organised by fellow members of the Oxford and District ARS. Much of the time was spent examining the various examples of mobile equipment, seeing how others installed mobile gear and looking at the various aerials – anything from 12ft centre-loaded whips to a 4-element Yagi for 2m. G3WW brought along his ex-New Zealand Government ZC1 that was mounted in the boot but remotely controlled from the driver's seat. G6AG had equipment for 2m, an EF91 crystal oscillator followed by a 6F17, 5763 and QQV03/20 with 25W input. Modulation was provided by a

The crowded scene in the main hall of the White Rose Rally 1974.

The National Hamfest 2011. when RSGB President Dave Wilson M0OBW greets ARRL representative Bob Inderbitzen NQ1R.

29th Longleat Rally.

Elvaston Castle Rally June 1993.

crystal microphone, 12AX7, 12AU7 and push-pull 6BW6s. The receiver used a 6AM4 RF stage and was a double conversion superhet.

Over the years, mobile rallies have changed slightly. No longer a gathering of operators showing their new mobile installation to other like-minded operators, they are now more like exhibitions that showcase the latest amateur radio equipment – mobile and fixed. One thing that hasn't changed much is the Bring & Buy where second-hand bargains are sought.

TELEPHONY CONTEST

The first ever RSGB telephony only contest was held in November 1956. The aim was to encourage stations to operate on the 21 and 28MHz bands during the years of high sunspot activity. For the first time, contacts between any station in the UK and any station in the rest of the world would count for points – 5 points for each contact and a bonus for each new country. More than 300 UK stations took part and between them worked more than 80 different countries. The leading station in the high power section was D Edwards G3DO who made 269 contacts with 101 of them on 21MHz and 168 on 28MHz. Of all these contacts, 100 were for a new country on a new band. The runner-up, James Taylor GM2DBX, worked even more stations – 303 – but had fewer bonus points. G3DO's station was impressive and comprised an Eddystone 888 receiver, RSC AR88 receiver, Labgear LG300 transmitter, speech amplifier, a G424 Panda Minibeam and a V beam with 310ft wire on each leg. The low power section was won by T Higginson GW3AHN. Using 25W with a 68ft Windom, he made over 140 contacts.

INTERNATIONAL GEOPHYSICAL YEAR

Amateur radio got involved in some radio research for the International Geophysical Year in 1957. Scientists throughout the world carried out an intensive programme of observation of various related phenomena occurring on the sun, in the atmosphere and on the earth. Radio amateurs in the UK were encouraged to contribute to the research by recording their observation. The first study in the UK was of aurora and related transmission condition on HF amateur bands between the UK and Canada. The next was a study of the relationship between VHF/UHF propagation and meteorological conditions. Then there was a study of auroral communication on the VHF amateur bands and finally a reception observation of solar noise in the VHF/UHF bands. Dr Smith-Rose coordinated the observations made by RSGB Members and forms were created to make sure the observations were consistent. Area Activity Coordinators were appointed and they passed the local research onto Dr Smith-Rose.

The launch of the USSR Earth Satellite, *Sputnik 1*, was part of International Geophysical Year. It carried two transmitters capable of 20.005 and 40.01MHz transmissions. Two weeks before the launch, the Royal Society asked RSGB members to provide several Doppler tracking groups for the first 36 hours, to record field strengths, fading rates and telemetry and to listen for each beyond the horizon signals. Seven RSGB Members' stations were chosen, spread around the UK; G8FC (RAF Locking), G3ENY (Bridgnorth), G2FNI (Wallasey), GM3EGW (Dunfermline), G3FBA (Bath), G3GDR (Watford) and G5BD (Mablethorpe). These stations worked in conjunction with the British Astronomical Association. For

D Edwards G3DO won the high power section of the first telephony contest held in November 1956 with this impressive rotary beam on his 45ft mast.

5. Post War to Century's End

two days, very good Doppler results were collected and sent to Norwood Technical College and a BAA group at Buckingham. Signals from *Sputnik 1* were reported as being received round the world in both directions at once with a resultant heterodyne between the two signals. This was because the differing relative velocity gave rise to different Doppler shift frequencies. It was probably the biggest thing that happened during International Geophysical Year, even though the suddenness of its appearance precluded much organised listening of its much publicised 'bleep'.

LIMITED 52MHz TRIALS

Late in 1957, following discussions between the RSGB and the GPO, it was agreed that selected amateurs in a few areas of the country would be permitted to operate on 52.5MHz, not exceeding 500W, for six months. The areas to be allowed that privilege were Northumberland, Monmouthshire, Glamorgan, parts of the Western highlands and Islands of Scotland. Other amateurs would only be allowed access to 52.5MHz, with special permission, between the hours of 0100GMT and 0930GMT. It is important to remember that, at this time, this frequency was in the middle of the television band and amateur use of such a frequency during television hours would be liable to cause severe interference. Twelve amateurs immediately requested permission and their names were sent to the Radio Services Department of the GPO for approval. More permissions were granted during the six month trial.

TECHNICAL TOPICS

The *RSGB Bulletin* of April 1958 contained a legend in the making. It was the first appearance of Technical Topics written by Pat Hawker G3VA. The new column aimed to "survey a few ideas from the amateur radio press of the world; a few hints and tips that have come to our notice, with perhaps an occasional comment thrown in for good measure". Pat wrote the column for the next 50 years until he retired with the

Sputnik, courtesy NASA

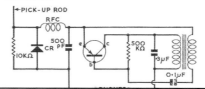

The first Technical Topics

Pat Hawker, MBE, G3VA

It was during his school days that Pat first became interested in radio and started building crystal sets, followed by a simple 1-valve audio amplifier. By 1935, he had progressed to a 2-valve short wave TRF receiver. Using this radio he heard many broadcast stations and, finally, his first radio amateur – LA1G. After that it wasn't long before he discovered the 40m amateur band and listened to British stations talking to each other.

After starting a radio club at school, he persuaded his father to take out an Artificial Aerial licence for him – as he was under 21, he couldn't hold a licence in his own name. His first callsign, in October 1938, was 2BUH. Once he passed his Morse test he received the callsign G3VA and it was around this time that he started writing for technical magazines of the day – including the T & R Bulletin (Simple Frequency Meter, 1939).

Like other radio amateurs in Britain, GPO engineers removed Pat's transmitter at the outbreak of WWII and G3VA went off the air for the duration of the war. The following year he received a letter from Lord Sandhurst asking if he would be prepared to so some voluntary work for the government. On signing the Official Secrets Act, he joined the ranks of the Voluntary Interceptors (VI). He was one of around 2,000 other radio amateurs who were approached during the war. Voluntary Interceptors were part of the Radio Security Service (RSS) and their original task was to listen for German agents in Britain. Later the VIs were tasked with logging Abwehr stations, the German Secret Service.

Pat was invited to join the full-time intercept stations and his Morse speed was tested to decide which job he should be assigned. He passed at the highest level – 25wpm – and was told to report to Hanslope Park as an intercept operator. In this role his bay contained two HRO receivers and each operator was given a list of signals to listen for. In 1944, Pat joined a new unit, SCU9, a small mobile unit headed to Normandy after D-Day. Pat was in Normandy until August 1944 before being sent to other parts of Europe as communications needs dictated.

In 1947, he joined the RSGB staff and learnt much about technical writing, going to work for George Newnes Ltd in 1951. He worked on various technical books including Radio & Television Servicing and transferred to Electronics Weekly in the 1960s. He joined the Engineering Information Service of the Independent Television Authority in 1968, staying there until his retirement in 1987.

Pat Hawker MBE G3VA.

Main party of the 18-strong Special Communications Unit No. 9 (SCU9) with 12 Section VIIIP operators and driver/mechanic, driver and two despatch riders in 1944. This was taken just prior to leaving for Normandy to provide a link between SIS and the main Army Command. Pat is third from the left in the back row.

April 2008 column being his farewell. For the first 10 years it appeared bi-monthly, then monthly for the remaining 40 years – some 600 columns and numerous ideas. His blend of clippings, contributed material and commentary was unique and was enriched by a lifetime's fascination with the technical aspects of radio. Pat was made an Honorary Life Vice President of the RSGB in 2008 and became a silent key in 2013.

VHF MATTERS

In October 1949, a plan for the better use of the 2m bands, suggested by G3CYY and G2XC in *Short Wave Magazine*, was introduced and was soon adopted by many VHF operators. The plan divided the UK into zones, each was allocated a certain sub-band of frequencies within the general limits of 144 to 146MHz. The object was to restrict the frequency range to be searched when the receiving aerial was pointed in any given direction and to facilitate contacts with DX stations by separating them from local stations. Following a meeting in 1958 between the Post Office, the Air Ministry and the RSGB, the 145-146MHz section was made exclusively amateur, provided changes were made in the band to discourage the use of the lower half of the band in those parts of the country where it was considered interference between amateurs and aircraft would most likely occur. A new plan was drawn up and approved and whilst there was some inconvenience to some operators, both the RSGB and *Short Wave Magazine* offered a crystal exchange service so that amateurs could swap crystals to enable them to end up with the right channels for their area. So, for example, Zone 1 (Cornwall, Devon and Somerset) was allocated 144.0-144.1MHz and Zone 2 (Berkshire, Dorset, Hampshire, Wiltshire, Channel Islands) was 144.1-144.25MHz.

The VHF beacon GB3VHF took to the air in 1960 from a BBC site at Wrotham. It was mounted on the mast at the 200ft level beaming north-west. The aerial array, donated by J-Beam Aerials Ltd, was a 5-element Yagi specially designed to withstand the severe environmental conditions at the site. The feeder was 300ft of 72Ω low loss heavy duty cable. The transmitter was a Plessey Type PT15A rated at 50W output located in a hut near the base of the mast. When measured, the frequency was found to be 144.4994MHz, which was a testament to the crystal grinding skills of G2UJ. It only varied by less than 1kHz at any stage. The beacon was in operation from 0600 to 2359GMT daily, although it did appear for 24 hours when tests were being carried out.

HF RTTY PERMITTED

In 1959, the GPO decided to allow RTTY on all amateur bands excluding 1.8-2MHz to authorised amateurs on a 12 month trial basis. The GPO also gave permission for FAX signals on 420MHz and higher frequency bands.

FIRST IARU REGION 1 CONFERENCE

The RSGB acted as host society at the IARU Region 1 Conference held in Folkestone in June 1960. A number of technical papers were presented, two of which were on RTTY and narrow band image transmissions, some of the latest ideas to capture the imaginations of European radio amateurs. Some of the decisions from this meeting urged Member Societies to encourage their members to make full use of all the amateur bands, for example, the HF segments of the 21 and 28MHz bands. All delegates were agreed on the importance of continuing to report the presence of persistent intruders in exclusive amateur bands. It was also agreed that the Executive Committee would look at the possibility of organising a European Foxhunting Championship. Looking at the new technology of data, Member Societies were encouraged to seek authority from their respective licensing

The IARU Region 1 Conference held in Folkestone.

authorities for interested amateurs to use teleprinting and to use the International Code No. 2 with an 850Hz shift for Frequency Shift Keying. It was at this meeting that the QRA Locator system was generally adopted by all amateurs in Region 1. The final document with all the details from the Conference ran to more than 100 pages of typescript.

AMATEUR SATELLITES

Was it coincidence or design that led to the satellite carrying amateur radio (Project OSCAR – Orbital Satellite Carrying Amateur Radio) going into orbit exactly 60 years to the day after Marconi and his colleague Kemp at Signal Hill, Newfoundland, heard for the first time a wireless signal that had originated from the other side of the Atlantic? The tiny capsule containing a transmitter operating on a frequency close to 145MHz orbited the earth every 91 minutes and provided excitement and interest unparalleled in amateur radio circles since the transatlantic tests nearly 40 years earlier. Reception reports reached the RSGB within a few hours of the launch. By the time the ARRL opened for business on 13 December 1961, a telegram of congratulations had reached them from the RSGB, as well as one to the Project Oscar Centre in California. Project Oscar was sponsored by the ARRL and it was thanks to the cooperation of the US Government that the transmitter was given space aboard a research satellite. The transmitted signal was 'HI' in CW. In the UK, the first report of the signals came from G3OSS who heard them at 0055GMT on 13 December. The Doppler shift of frequency due to the high velocity of the satellite was measured by G3GDR as about 7kHz. Details of the signals were passed on to W1FRR on 80m SSB, who relayed the message to the Project Oscar Association in the US. Sadly, the satellite failed earlier than was expected but not before there was time to complete most of the main objectives of the investigation. The last proper signal was heard early on 30 December. By 1116GMT, the keying system had failed and only 'II' was being transmitted. A number of stations heard a continuous note on 144.96MHz at 2305GMT. The strain on the batteries was just too much and the power had reduced to a small fraction of the original 100mW. It continued with this warbling note until 3 January at weaker and weaker levels. At the end of the experiment more than 2,500 reports had been sent in from around the world. Reception distances of around 1,400 miles maximum were reported.

Australis-Oscar 5 was successfully launched on 23 January 1970 at 1131GMT. It was first heard at around 1243GMT in the UK with good signals from both the VHF and HF transmitters being reported. It was the first amateur satellite to operate several telemetry channels and capable of being controlled from a ground station. It was possible to copy nine consecutive orbits each day from around 0100 to 1700GMT. Amateurs were asked to keep clear of 144.050MHz during orbit times to allow listeners to hear the satellite. The HF beacon was transmitting on 29.450MHz. Bill Browning G2AOX was kept busy providing orbit prediction for amateurs in the UK.

The World Administrative Radio Conference (WARC) for Space Telecommunication took place in June 1971 and agreed to proposals to allow the amateur satellite service allocation in the 7, 14, 21, 28, 144 and 435MHz bands as well as 24GHz.

Bert Allen G2UJ (left) and Angus McKenzie G3OSS examining a model of OSCAR-1. Angus was the first UK amateur to report reception from OSCAR-1 as it orbited the earth.

The flight prototype U-band patch array under test on the Phase 3-D integration facility antenna range. **Photo by KB1SF.**

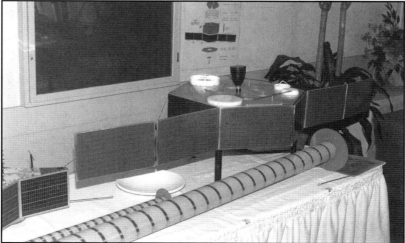

Flight model prototype of Phase 3-D's S-band antenna.

VHF NATIONAL FIELD DAY

The first VHF National Field Day took place in July 1962. There were 39 entries with 139 operators involved. The winners by a clear margin were the Wolverhampton Group GW3KMT/P operating from Denbighshire near Oswestry. Second was the North West VHF Group G3OHF/P operating from near Leek in Staffordshire. In the first contest, scoring was complicated and double points were awarded for working a mobile station. The adjudicators found that many entrants lost points due to logging errors.

The 1963 event was moved to September and became a multiband portable contest (70, 144, 432 and 1296MHz). 45 clubs entered with over half of them running stations on 2 bands. Overall winners were Surrey Radio Contact Club running G2RD/P and G3ODY/P and second place were Crawley ARC running G8RW/P and G3FRV/P.

Dorking & District ARS had their B station on Leith Hill, Surrey, at 997ft ASL. For 70cm a corner reflector built by G3OVS gave good results, whilst for 70MHz a 4-element Yagi was used.

Durham City ARS 4m station G3UTS/P. G3UIR and G3UIW operating.

5. Post War to Century's End

48-element 70cm antenna and 23cm parabolic dish as used by G3TND/P in 1967.

Purley & District RC 3cm station G3WZR/P on Kenley Common in 1968.

Leyton & Walthamstow Group. **Photo by G2HR**

GOLDEN JUBILEE

In 1963, the Society celebrated its Golden Jubilee and a week long programme of activities culminated with a reception and dinner at the Connaught Rooms, where the principal speaker was Lord Brabazon of Tara. Messages of congratulations were received from many parts of the world with comments such as "remember life begins at fifty", "continued success for the future" and "I have been probing my old crystal set with a cat's whisker, found several bright spots all forecasting continued success for the RSGB". The latter comment was from Constance Hall G8LY. Jack Hum G5UM said, "few of us are likely to see the next Golden Jubilee of the RSGB so let's make the most of this one". From the reports and letters following the celebrations, the amateurs who participated certainly took Jack's sentiments to heart. At the BBC they were given a behind the scenes view of everything from the props workshop to the control rooms and studios. The visit to the Radio Research Station inspected the project investigating the lower ionosphere by rockets and an investigation of atmospheric noise and the results obtained from data collected during the International Geophysical Year. Later that evening Mullard Ltd held an open house and members had the chance to see something of the latest advances being made by the company. A large party of members boarded the *Marchioness* for a day long trip to Hampton Court Palace on 4 July. During the voyage, GB3RS/M, specially authorised to operate on the tidal section of the Thames, was active using the new KW Electronics SSB transceiver on a 14MHz dipole. The celebrations culminated on 5 July with the Golden Jubilee Dinner. It was attended by more than 400 Members, friends and overseas visitors and was described as, "all the Conventions and Regional Meetings rolled into one to the 'nth power".

RSGB Golden Jubilee Dinner

LICENCE CHANGES

On 1 June 1964, amateur licensing underwent a huge change. The new licences were Amateur (Sound) Licence A, the Amateur (Sound) Licence B – a new phone-only UHF licence – the Amateur (Sound Mobile) Licence and the Amateur (Television) Licence. This new B licence was restricted to the use of frequencies above 420MHz and was phone-only. The callsigns issued were in the G8 plus three letters series and required passing of the Radio Amateur Exam only and no Morse test. From that point, the amateur television licences were G6 /T callsigns. Other changes involved the repetition of the callsign, log keeping, guest operators and emissions types. This was the first time that UK radio amateurs could attain a licence without having to pass a Morse test. The Licence B remained unchanged for four years when the 144MHz band was added to the permitted bands.

Big news appeared in the *London Gazette* in May 1987. From 1 June 1987, Class B licensees would have access to the 50 and 70MHz amateur bands. Both bands had an increased allocation from that time too. Amateurs had primary status on the 50 to 51MHz section and secondary status from 51 to 52MHz. Operation /A and /P was permitted but /M was still not allowed. The only other restriction was that 50MHz antennas had to be below 20m AGL and polarisation was to remain horizontal to protect television broadcast transmitters in Europe. The 70MHz band was expanded to 70.0 to 70.5MHz with amateurs having secondary status.

ISLANDS ON THE AIR

Created in 1964 by Geoff Watts and taken on by the RSGB in 1985, Islands On The Air (IOTA) is an amateur radio activity programme designed to encourage contacts with island stations worldwide. The islands of the world have been grouped into some 1200 IOTA groups with, for reasons of geography, varying numbers of 'counters' or qualifying islands in each group. The objective, for the island chaser, is to make radio contact with at least one counter in as many of the groups as possible. For the DXpeditioner, the object is to provide such island contacts. A variety of certificates and awards are available for this programme.

Ken G3OCA & Cezar VE3LYC during their activation of two very rare Labrador IOTAs. They got to the remote island by chartered helicopter with all their equipment.

MILESTONE PUBLICATIONS

An article appeared in the November 1966 *RSGB Bulletin* that will be familiar to many amateurs, even today. The G5RV Aerial – Some Notes on Theory and Operation by Louis Varney G5RV described a multi-band dipole specifically designed with dimensions that allowed it to be installed in most normal sized back gardens, giving effective operation from 1.8 to 30MHz. Full constructional details were given and you have to wonder just how many of these aerials have been built made over the years.

For the first time, the history and highlights of amateur radio spanning more than 50 years was available as a book. *World at their Fingertips* by John Clarricoates G6CL, was written when many of those early pioneers were still living and the book gives an account of the British amateurs part in the development of the hobby.

CONSTRUCTION UNDER THREAT

The Wireless Telegraphy Bill in 1967 caused amateurs some concern at it was going through Parliament. Clause 7 of Part II gave the Post Master General the power to prohibit manufacture or construction of any type of radio apparatus whether or not for sale. The PMG stated in Parliament that the desire to prohibit manufacture and importation of radio equipment was aimed at 27MHz walkie-talkie and 'bugging' devices. Unfortunately, the Bill's terms could include almost any form of radio equipment deemed by the PMG to be undesirable. Mr G D Wallace MP was willing to speak on the Society's behalf when the Bill was read a second time in April. He put forward the arguments on behalf of the amateur radio movement. Subsequently, Mr P Bryan MP, the opposition spokesman on Broadcasting and Allied Matters, invited R F Stevens, G2BVN, RSGB President, to attend a meeting of the Conservative Broadcasting Committee to explain the views of the Society. Both MPs tabled amendments that were considered. Whilst these amendments were turned down, the RSGB did get reassurance from the PMG that "the Post Office will continue its close contact with the RSGB, the amateurs' representative body. Before I make any orders specifying apparatus which the manufacture is to be banned, I will discuss it with this body". This settled concerns that this Bill would be used to the detriment of the amateur service.

FESTIVAL OF LONDON.

GB2LO was an RSGB exhibition station at the 1968 City of London Festival. Finding a suitable location proved difficult as it needed to be in a building with public access and where a good aerial could be erected. The *Daily Mirror* building fulfilled all these requirements and, due to good relations with the newspaper, the owners of the *Daily Mirror* agreed to help with the installation. The station was on the air from 8 to 20 July and open to the public from 11am to 4pm. Members could operate the station if they brought their licence with them. The *Daily Mirror* obtained permission to mount an aerial on the roof and to erect a temporary building on the pavement at Holburn Circus. The newspaper then went on to erect the building, carpet and furnish it, install a telephone and provided the handouts and QSL cards. The aerial was a 3-band 2-element quad. The operators were experienced SSB HF operators able to pick out the right kind of signals that would attract the attention of the public. The RSGB had decided to use commercial equipment rather than showcase homebrew projects, which worked fine because very few visitors paid much attention to the equipment. The KW equipment worked well throughout the 13 days of operation during which they worked 108 different countries. During public hours a limited number of

5. Post War to Century's End

Festival of London GB2LO.

Rowley Shears G8KW with the equipment used at GB2LO.

contacts were made as the contacts were carefully chosen to showcase amateur radio. After public hours they changed to more contest style operating. The choice of callsign with its sentimental associations of early BBC days was successful and the BBC covered the event in several radio and television programmes – one of them in Finnish.

23cm RECORDS

On 22 March 1970, the Cambridge University Wireless Society DXpedition to the Isle of Man worked GW3XAD/P on the 1296MHz band. It is thought that was the first GD to GW 23cm contact. Operating as GB3TPF/P, they worked G3XAD/P the following day using CW this time, and this is believed to be the first GD to G 23cm contact. Both contacts were at the 100 miles limit but were made all the more difficult due to the weather. G3TPF reported that, "keeping the dish pointing in the right direction with 100mph gales to contend with was not easy".

TECHNICAL LECTURES

The RSGB Education Committee delivered the Christmas Lecture for young people at the Science Museum on 6 January 1973 – the first of several years when this took place. The lecture was given in two sessions, each lasting about 1½ hours. Aimed at 6th formers with a casual interest in

Norman Kendrick G3CSG showed how RF can be indicated on a field strength meter, the results were shown on a closed circuit TV for the audience to view.

Len Newnham G6NZ at the Science Museum lecture.

radio, Len Newnham G6NZ spoke about the reasons you would get involved in amateur radio. This was followed by Norman Kendrick G3CSG who explained how we communicate - he used 14 different demonstrations including a closed circuit TV broadcast to magnify the field strength meter and other small components so that everyone could see. John Swinnerton G2YS explained all about single side band and the RSGB HQ station, GB3RS, took part in a live QSO as part of a session on how to get started in the hobby. This was alongside a set of slides showing a wide variety of amateur stations. Finally, Ron Wallwork G3JNK explained all about QSL cards and gave each of the audience an SWL card to send to HQ. At the end of the two sessions, students visited the Science Museum station, GB2SM, for a second demonstration, this time working some DX with full power and a rotary beam.

REPEATERS & BEACONS

The Pye Telecommunications ARC built the first UK repeater, GB3PI, in Cambridge in 1971. A temporary experimental licence was granted and the repeater was established, initially, at the Pye Telecommunications premises where permanent security staff could, if necessary, switch the unit off at any time. It was designed to transmit on 145.75MHz with just 10W with the receive frequency of 145.15MHz. The opening tone needed was a 1,700Hz tone for half a second and it also had a 1 minute timer in case the repeater was held open inadvertently. The group was awarded the Courtney Price Trophy for outstanding technical development in the field of amateur radio in connection with the GB3PT repeater at the 1973 AGM. After 18 months of operation, the group replaced the original unit with one that had new operating procedures to allow more people to use the repeater.

The first 10GHz beacon was established on the Isle of Wight on 3 April 1975. Operating continuously on 10.100GHz it had an RF output of 80mW to a horizontally polarised omnidirectional aerial with around 10dB gain in the vertical plane – an ERP of 0.8W. The beacon was located on St Catherine's Hill at 800ft ASL and was housed in a small concrete building belonging to Pye Telecommunications Ltd. Prior to installing the system, those behind the beacon had put it through a 12-week non-stop bench test during which time no faults developed. The transmitter was a Plessey Gunn oscillator with varactor electronic tuning, Type GVDO-101, a 30dB isolator, a 20dB directive coupler (for the wavemeter) and a 30ft long waveguide run between the bench and the aerial. The transmitter was frequency modulated by superimposing a 700Hz sinewave on the varactor's DC potential. Keying was accomplished by a 128 bit digital keyer that transmitted the callsign over a 12 second long period, followed by a 10 second long dash.

NOBEL PRIZE WINNERS

The Society offered its congratulations to Professor Sir Martin Ryle, FRS, G3CY, the Astronomer Royal, who jointly with Professor Antony Hewish was awarded the 1974 Nobel Prize for Physics. Both men had worked together at the Cavendish Laboratory, Cambridge for 25 years and were awarded the prize for their pioneering work in radio astronomy. Professor Ryle, who had been Professor of Radio Astronomy at Cambridge since 1959 and was Director of the Mullard Radio Observatory, was an Honorary Member of the RSGB.

10GHz tests over a 114 miles sea path when G3RPE/P near Cromer worked G3ZGO/M on Flamborough Head on 20 June 1973.

MOBILE LOGS

The Home Office agreed with a proposal from the RSGB to simplify mobile logging in 1975. The details were limited to; date, time of starting and finishing the journey, starting and finishing points of the journey and frequency band(s) used. It was a welcome concession and did much to make mobile log keeping manageable.

MULTIPLE CHOICE RAE

From 1979, the Radio Amateurs' Exam (RAE) took the form of multiple choice questions. The City & Guilds asked the RSGB for help in testing the questions by conducting sample exams on candidates who had reached examination standard. Members of the RSGB Education Committee were at the Amateur Radio Retailers' Association (ARRA) Exhibition in Granby Halls, Leicester, to meet RAE instructors and lecturers to discuss the new syllabus and the changeover to multiple choice questions.

WARC

The 1979 World Administrative Radio Conference (WARC) comprised 74 days of meetings, 2000 delegates/observers from 142 countries and 30 international organisations, 894 plenary meetings or meetings of committees and working groups, plus many smaller meetings. The new Radio Regulations agreed at the Conference would come into effect on 1 January 1982. Three new amateur bands were announced at 10.1, 18.068 and 24.890MHz. In the UK, 10MHz did become available on 1 January 1982, but the other two bands remained allocated to the fixed and land mobile services until assignments were transferred to new allocations. It wasn't until mid-1989 that the latter two bands became fully available to UK radio amateurs. Another announcement was that the frequency above which administrations could issue licences without a CW test was given as 30MHz. Finally, new amateur satellite segments were noted for bands between 7 and 240GHz.

HM KING HUSSEIN OF JORDAN JY1

RSGB Member HM King Hussein of Jordan JY1 made a short visit to the UK in late January 1980. In a letter to the RSGB General Manager, he conveyed best wishes to all RSGB Members. JY1 was operational while in the UK using his UK callsign G5ATM.

King Hussein of Jordan JY1 was interviewed by *RadCom* in June 1993 at the Jordanian Embassy in London. He explained what interested him about amateur radio, "I've enjoyed talking to all age groups with all kinds of interests". He went on to describe his contact with the first amateur in space, Dr Owen Garriott W5LFL. "We managed to arrange a schedule with him on his 92nd orbit. It was an excellent contact, something like three or four minutes horizon to horizon. We were following him with a directional aerial". JY1 became a silent key in February 1999.

AMATEUR SATELLITES

The second test start of the ESA Ariane rocket carrying the Amateur Phase 3A satellite was scheduled for 23 May 1980. However, after around 58 seconds the computer interrupted the countdown. A second countdown was started but was also interrupted about 53 seconds before the planned start. A tropical storm caused a further delay but, finally, the Ariane rocket lifted off from the launch pad at 1429UTC. A report of an alteration in pressure in one of the four Viking-V motors was noticed and between 60 and 100 seconds after launch, this motor failed completely. The rocket went off course and Ariane fell into the sea some 600km from the launch site and the amateur satellite was lost.

The following year, UoSAT 1 was constructed at the University of Surrey, supported by the RSGB and AMSAT. It carried four phase-related HF beacons on 7, 14, 21 and 28MHz to provide tools for propagation studies, a 450mW beacon operating on 145.825MHz, a 400mW beacon on 435.025MHz and beacons on the 2.4 and 10.5GHz bands.

HM King Hussein of Jordan JY1

The satellite exceeded the expected two year lifespan by six years, with the final signals on 13 October 1989 when it re-entered the atmosphere.

AMSAT-UK, the amateur satellite organisation of the UK, made a donation of £32,000 to the AMSAT Phase 3D satellite fund in 1993. The RSGB gave AMSAT-UK a £1,000 donation in 1993 and £25,000 in 1995 for the Phase 3D project. The Phase 3D satellite had seven uplink and six downlink bands between 15m and 1.5m and offered the capability of worldwide QSOs to an enormous number of amateurs because of its modest demands on the ground stations, both in regards to antenna gain and transmitter power. Because it was designed for an elliptical orbit of about 16 hours period, manual antenna tracking was perfectly adequate for most of the orbit.

Even with the donations from the RSGB and AMSAT-UK, the project was still short of the total amount required to pay for the launch. Eventually the launch was underwritten when other donations to the project were received. It was launched on 16 November 2000.

NOISE & INTERFERENCE

Concern about interference from mains-using devices started in late 1980. An increasing number of devices were being developed that used the consumers' electrical wiring installation as a communication or signalling channel. This led to a revision in the Electrical Recommendation (G22/1) covering superimposed signals in the public electricity supply network. The devices included speed communication systems and remote alarms. The potential problems were identified as the distortion of mains voltage waveform that could affect consumers' supply; the injection of signals into the supply network that would cause interference with other communications equipment; the provision of a blocking filter to limit the injection of signals that could interfere with the supply control signals from the area electricity board; and the electrical safety of such devices and the possibility that the electricity board and consumer signalling would interact with each other. The Post Office was concerned about the possibility of radio interference from these devices and said that there might be a need to "consult the Home Office regarding licensing arrangements".

In June 1998, interference via the mains raised its head again. Details appeared in the press about the use of the electricity mains to distribute high speed data signals for the internet and similar services. The RSGB EMC Committee took part in meetings chaired by the Radiocommunica-

UoSAT-1 team at Vandenberg Air Force Base. **Courtesy AMSAT-UK**

tions Agency, at which the company proposing the systems as well as the users of the HF bands were present. The EMC Committee put forward the argument that there was a need to protect the amateur bands and other vulnerable services from interference. In 1999, the RSGB EME Committee made prominent politicians and the national media aware of its concern about Power Line Telecommunications or PLT. They issued a combined press release and briefing pack pointing out the concerns of radio amateurs and other radio users.

In 2001, the RSGB voiced its concerns again on mains interference when the German Government approved the introduction of Power Line Communication (PLC) systems as long as they complied with regulation NB30. The move was strongly opposed by radio users in Germany, which was supported by the RSGB because "The emission levels in NB30 are much higher than the levels proposed by UK HF radio users as a 'worst case' for acceptable interference from such systems". The RSGB reiterated its commitment to "fight for the preservation of the radio spectrum as a valuable natural resource, and is making its position known through all available channels, both nationally and internationally".

MODERN TRANSATLANTIC FIRSTS

The first transatlantic 70 to 50MHz QSO took place on 17 November 1980 at 1627UTC between G4BPY and VE1ASJ. The received signal from Canada was 5 and 9 while the CW from the UK was 339 in Canada.

Paul Turner G4IJE made what was probably the first transatlantic 50MHz SSTV contact at 1320UTC on 22 November 1989. He received a picture from Dave Faucher WA1UQC on 50.23MHz. Dave was running 300W to a 6-element Yagi at 50ft and a Robot 1200C Scan Converter. Paul used a TS-660 and homebrew RGB framestore. All the SSTV signal processing was done via a BBC Micro. Two-way SSTV also took place over the following two days between WA1UQA and G3NOX, G1LXI and G4IJE.

SECOND UK IARU REGION 1 CONFERENCE

The 12th triennial IARU Region 1 Conference was hosted by the RSGB in Brighton in April 1981. National Societies from 38 countries were in attendance. The Conference was opened by Rt Hon Timothy Raison, MP, Minister of Home Affairs. The first call for papers was made in June 1980 and 144 papers were prepared for the various meetings. Band plan proposals for the new WARC bands were put forward ready for when the bands would come into operation. It was agreed that the transmitter power on 10MHz should not exceed 250W mean output power, no contests should be organised on the band and that SSB should only be used during emergencies involving immediate safety of life. The band plans for other bands were discussed and minor amendments were agreed. One of the many results from the conference was that the dividing line between UHF and microwaves was confirmed as 1GHz because "operation above this frequency was still highly experimental".

IARU Region 1 Conference

IARU Region 1 Conference.

The conference station, GB1IAR, was active on all bands, HF and VHF, and made more than 5,000 QSOs. The station was popular with delegates, particularly late in the day. Twelve Sussex radio clubs planned the station and they negotiated the loan of all the equipment from radio dealers. There was an FT-101E plus Magnum 4 transverter and 5-element beam for 70MHz. The 144MHz SSB and CW station used a TS-700S plus 100W linear and 8-element Tonna, the FM station was run using an FT-221R putting 15W into a collinear. UHF activity was achieved on SSB and CW using a Multi 750E and 88-element beam with FM from a Philips FM321 both into a collinear. Home-made equipment from the Sussex Repeater Group was used on 1.3GHz. The Top Band station was a Drake TR7 with a dipole and 80m operation used a TS-820S running 200W PEP into an inverted V. A TS-130S running 200W PEP into another inverted V was used for 7MHz with an IC-720A plus 2KL linear to a 3-element beam for 14MHz. The 21MHz position used a TS-830S into a 3-element beam and an FT-101E plus YC601 digital display saw 200W into a quarter wave vertical for 28MHz. Finally, RTTY on mainly 14 and 144MHz had a CT100 terminal unit, FT-707 and filter. At no time during the Conference were there less than three stations in use and, as time headed towards midnight, there were often 10 of the stations operating simultaneously.

FALKLANDS WAR

On 2 April 1982, Argentina started to make claims that it had invaded and taken over the Falkland Islands. They seized control of the international telephone and radio networks on the islands and so the Governor, Rex Hunt, was unable to contact the Foreign & Commonwealth Office to report his surrender to the Argentine Forces. Although the Argentine dictator General Galtieri was proclaiming that "the will of the people has been fulfilled", the British Government had no independent verification of the Argentine claims.

Les Hamilton GM3ITN had been having regular contacts with Tony Pole-Evans VP8HZ on Saunders Island, just off West Falkland since the 1960s. He also regularly spoke to Bob McLeod VP8LP at Goose Green. He heard from VP8LP that the Argentine flag was flying over Government House in Stanley and when Bob went off the air Les realised that the Argentines had reached Goose Green.

Les was the first person outside Argentina and the Falklands to know that the invasion had taken place. He phoned the Ministry of Defence, who informed Margaret Thatcher, the Prime Minister. British Military Intelligence asked Les to keep in contact with the Falkland amateurs as much as possible and in the weeks that followed he was able to pass on reports from Tony Pole-Evans that were of vital importance in the liberation of the islands.

NFD WEATHER

June 1982 went down in the record books as the wettest month of the century in many places. Typically, the early summer weather broke in time for HF NFD. Of the 100 plus stations that entered, 14 were unable to finish because of station equipment damage caused by the severe storms that swept across much of the UK. Apart from thunderstorms, some areas experience gale force winds and monsoon-like rains. Sixty one stations experienced some kind of problem, 42 of them had to close down for a time and not all were able to get back on the air. Many sites had near or direct lightning strikes that took out the front ends of transceivers, welded coax, burnt antenna tuning and switching units and damaged power supplies and other accessories. On one site, the FT-101ZD caught fire, on another the generator was struck and two operators suffered minor burns when their station antenna was struck. The NFD trophy was won by Racal Amateur Radio Group G3KLH/P with 3,455 points as the Open Section winner. The Restricted Section winner, and winner of the Bristol Trophy, was the Great Western Group G3NKS/P with a score of 3,185.

NFD wettest ever!

FIRST LADY PRESIDENT

The RSGB President for 1985 was elected at the Council meeting on 11 August 1984. Mrs Joan Heathershaw G4CHH had been a Council member for Zone A since January 1980 and was the executive Vice-President for 1984. She was the Council's first lady member and also the Society's first lady President. She was introduced to amateur radio by her husband, Duncan G3TLI, and, during her installation speech, she spoke of the enjoyment and satisfaction that amateur radio gave her. Her local radio club, Hornsea ARS, presented her with a hand-made silver-plated Morse key.

During her time in office, she visited around 40 different clubs as well as many trophy presentations. At the end of the year, she said that, "it has been a year of streamlining and improving efficiency of balancing continuing improvements against resources" and she looked forward to continuing her efforts on behalf of the RSGB in her position as immediate Past President. She was re-elected as President in 1987.

Joan after handing the RSGB Presidential chain over to Sir Richard Davies KCVO CBE G2XM in 1988.

Joan Heathershaw G4CHH.

AMATEUR RADIO IN SPACE

In 1983, one of the astronauts on board the STS-9 Space Shuttle was Dr Owen Garriott W5LFL and he obtained permission to take a 144MHz handheld on board running 5W into a printed circuit board antenna placed in the upper crew compartment window on the aft flight deck. During his free time, Garriott transmitted on the even minutes and received on the odd minutes. During the receive periods he logged all the callsigns he heard and then acknowledged them in the transmit minute. He recorded about 290 amateurs around the world during the flight, although his transmissions were logged on earth by many thousands more. Gloria Hills G4UYL, Jan Niven G6EGY, Sue Nelson GM8NXC and J W Hoyland G6DEF were among the five UK radio amateurs who were recorded as UK contacts. Of course, probably the highest profile contact was with RSGB Member JY1, King Hussein of Jordan. Those who managed a two-way contact received personally signed QSL cards.

Garriott's son, Richard W5KWQ, flew to the space station in 2008 and spoke to Budbroke School in Warwickshire. In 2009, Richard visited Budbroke School to meet the children he had spoken to whilst in space.

In late 2000, the crew from *Expedition 1*, KD5GSL and U5MIR, arrived on board the International Space Station. They were the first to operate the permanent ARISS station aboard that included VHF and UHF handhelds and a packet TNC. Initially, operation was just on 2m FM and packet but the plan was to move to multi-band multi-mode operations, including regularly scheduled school group contacts. Two US callsigns were issued; NA1SS for use on board the ISS and NN1SS for ground-based ISS communications from Goddard Space Flight Center in Maryland. A Russian callsign, RZ3DZR, and a German callsign. DL0ISS, were also issued.

5. Post War to Century's End

Dr Owen Garriott W5LFL was the first radio amateur in space

Richard Garriott W5KWQ visited Budbroke School.

Interior view of the Space Shuttle Orbiter Columbia's flight deck. **Courtesy NASA**

Owen W5LFL & Richard W5KWQ at the 2009 Dayton Hamvention where they were interviewed by *RadCom*.

YOUNG AMATEUR OF THE YEAR

The first Young Amateur of the Year award was issued in 1988, the RSGB's 75th Anniversary year and was sponsored by the Department of Trade and Industry (DTI). The DTI said that, "Anybody under the age of 18 who has made waves in the world of amateur radio should enter the Young Amateur of theYear award". It was designed to increase awareness of amateur radio amongst young people and highlight the skills and benefits that participation in amateur radio can bring. The first award was won by Andrew Keeble G1XYE. Andrew lived in Norwich and was nominated for his work within the Scout movement, RAYNET and his interest in the study of propagation and antennas. The presentation was made by HRH the Duke of Edinburgh at the RSGB's 75th Exhibition and Convention at the NEC.

In December 1997, the 1997 Young Amateur of the Year, Emma Constantine 2E1BVJ, featured on BBC TV's Blue Peter. Emma was interviewed by Richard Bacon and then went on to make contact with 2E1BOO and 2E0APH on the 70cm band. A special event callsign was organised to allow the presenter to pass a greetings message to both stations contacted. During the interview, Emma pointed out that amateur radio equipment did not need to be expensive and demonstrated a £5 crystal set and a £25 transceiver kit. Following the broadcast, the RSGB received more than 70 enquiries from young people and their parents who wanted to find out more about amateur radio.

75th ANNIVERSARY

In the 75th Anniversary year, His Royal Highness Prince Philip, Duke of Edinburgh – the RSGB's Patron since 1952 – opened the Convention at the NEC. He made the formal opening address, presented the Young Amateur of the Year Award and toured the displays. His opening address was

HRH the Duke of Edinburgh presented the first Young Amateur of the Year award to Andrew Keeble G1XYE.

carried in a special edition of GB2RS that was broadcast live from the NEC. He said, "It has been some time since my last appearance as Patron of the Radio Society of Great Britain. In fact it was in 1966 at the opening of the National Exhibition in London. Now this may look rather like a dereliction of duty but in my opinion Patrons should hover benignly over their charges rather like guardian angels and should only become involved on very special occasions and obviously the 75th Anniversary is just such an occasion and I'm delighted to manifest myself with this celebration. When I accepted the invitation I thought I'd just check up to see when I became Patron of this Society and I was astonished to find that I'd been hovering over it ever since 1952. My quick calculation reveals that this was 36 years ago and that's very nearly half the life-span of the Society.... I dare say that radio amateurs would have struggled along without the help of the Society but the hobby would never have reached the standards or the popularity that it has today. Furthermore the Society can be proud of its contribution to the development of this hobby world-wide and for the part it plays in the International Amateur Radio Union."

"1913 is quite a long time ago and I can just imagine the excitement it must have caused when the first amateurs discovered that wireless telegraphy really worked and that they could, I imagine, sometimes communicate with others at a considerable distance. Of course the equipment has vastly improved over the years and the chances of making contact with others are probably that much better but I suspect that the excitement and the sense of achievement is as great as ever. And that excitement must have been much enhanced when the 'ham' finds that he can put his skill to practical use in times of war or emergency...."

The RSGB Patron, HRH Prince Philip, Duke of Edinburgh visited the 75th Anniversary Convention.

AGM ON THE ROAD

In 1988, the RSGB held the first AGM outside of London at the University of Manchester Institute of Science and Technology. The first part of the meeting was the formal AGM where the Minutes from the 61st AGM and the Accounts for the year ending 30 June 1988 were approved. The second part of the meeting began with the presentation of awards by the RSGB President, Sir Richard Davis KCVO G2XM. After the awards, there was a 15-minute video of the official opening of the 75th Anniversary National Convention by HRH Prince Philip. The video covered the opening address and presentation of the DTI-sponsored Young Amateur of the Year Award to Andrew Keeble G1XYE. The video was another first for the Society and was the direct result of volunteer effort by members of the British Amateur Television Club.

The presidential address looked back 75 years to the early days of the Society and looked ahead to proposals for a new beginners licence. Later that evening, Sir Richards handed over the chain of office to Dr Julian Gannaway G3YGF, the Society's President for 1989.

RAYNET OPERATIONS

At the very end of 1989, RAYNET personnel in Scotland were called to assist in one of aviation's most tragic events – Lockerbie. Teams from Dumfries and Galloway and Strathclyde turned out and operations began at 2200UTC. One of the controllers recalled his first impressions, "There were rows and rows of ambulances, fire hoses were stretched across the roads and groups of people had gathered in the street. The smell of burning hung over the town". RAYNET set up communications links between Lockerbie and Dumfries, something they maintained for 10 days. The teams passed messages to and from the Emergency Planning Officer (EPO) in Dumfries and handled traffic for the Scottish Ambulance Service. They also set up several bases and provided operators for the search areas. In true amateur spirit, help from other groups in Scotland, England and Wales was immediately offered.

At times Scottish RAYNET provided communications for not only the EPO and police search teams but also for the Air Accident Investigation Unit, the Search & Rescue dogs and police dogs teams, the Salvation Army, the underwater search units, the Red Cross and the rescue helicopters. The DTI had cleared them to work the aeronautical stations as it wasn't covered in the amateur licence. The teams even worked two shifts on Christmas Day.

Altogether over 7,000 man hours were provided by around 80 operators each day. The teams finally stood down at 1720UTC on 31 December 1989. This was a truly remarkable effort in extremely difficult circumstances.

An MBE was awarded to Alexander Anderson GM4VIR for his work with RAYNET at Lockerbie. He had been the controller for the Dumfries & Galloway RAYNET Group since the formation in 1985. On receiving his award, Alex emphasised that he considered the award to be for all the many RAYNET Groups and members who helped in the operation.

FIRST BRITISH AMATEUR IN SPACE

Helen Sharman was selected to be the first British astronaut in 1991. To encourage schools to participate in experiments aboard the Mir Space Station, the DTI issued the special callsign GB1MIR to Helen and a series of callsigns with rare four letter suffices to the participating schools (GB0JUNO was issued to Harrogate Ladies College). Harrogate had six licensed operators amongst the school radio club ready to ask Helen questions during the 10 minute window for communications. Helen completed her six day mission on the Mir Space Station and worked on three British projects. One was

5. Post War to Century's End

taking high resolution pictures of the Earth for educational purposes, the second was bringing back seeds so that 5,000 schools could compare those that had travelled in space to those that had stayed on Earth and, finally, there was the amateur radio project to inspire schools. Towards the end of her mission, Helen had the opportunity to make random QSOs with several British amateurs. Helen was made an Honorary Member of the RSGB when she visited RSGB HQ in January 1993.

Helen Sharman GB1MIR.

Helen Sharman receives her Honorary Membership from RSGB President Peter Chadwick G3RZP.

Centenary

The first four Novice licensees – 2E0AAA, 2E1AAA, 2E1AAD and 2E1AAE all met the RSGB Patron HRH Prince Philip, Duke of Edinburgh, KG.

NOVICE LICENCE

The UK Novice Licence was introduced in 1991. It was conceived to enable more beginners of all ages, but especially the young, to try out amateur radio. Novice B licensees were permitted 3W output on frequencies above 30MHz with a licence that was designed to provide an initial incentive to get on the air. The Novice A licence that gave access to HF frequencies as well required the completion of a 5wpm Morse test.

The Novice licence had an agreed training course of instruction that aimed to provide a thorough grounding in the basic principles and concepts of amateur radio and to teach good operating skills. The course was around 30 hours long and, once completed satisfactorily, the student was able to apply to sit the multiple choice exam that was held every three months. Some 200 students completed the Novice training Course in time to take the very first Novice RAE on 3 June 1991. The Telecommunications Minister presented the first Novice Licence with its distinctive 2 callsign on 25 July. Jonathan Page 2E1AAA was the first Novice B licensee and Hugh McNeill 2E0AAA was the first holder of a Novice A licence. The first four Novice licensees – 2E0AAA, 2E1AAA, 2E1AAD and 2E7AAE - all visited Buckingham Palace in November to meet the RSGB Patron HRH Prince Philip, Duke of Edinburgh, KG.

Front row left-right: Hugh McNeil 2E0AAA, Simon Khan 2E1AAB.
Back row left to right: David Hull 2E0AAB, Jonathan Page 2E1AAA, John Redwood MP, Robert Cherry 2E1AAC, Nicky Foster 2E1AAD, Natasha Weir 2E1AAE.

GB0OSH

Justin Johnson G0KSC set up a scheme to raise money to help Great Ormond Street Hospital using amateur radio. He recruited the help of several Essex amateurs (G1OGY, G0PAE, G4ETG, G1EVD, G1BTF, G7ABL, G0BDC and G7CWX) to help stage the special event station, GB0OSH, in late February 1993. He was able to borrow all the necessary equipment and arranged to set up the amateur radio station inside the accommodation for the hospital broadcasting station.

The RSGB Patron, HRH Prince Philip, Duke of Edinburgh, KG was also invited to open the station. With RSGB President, Peter Chadwick G3RZP, looking on, Prince Philip passed a greetings message over 2m to Mike Wickham G4IGK. After the royal visit, some of the young patients had the opportunity to get on the air. Later in the four days of operation the BBC childrens' programme Blue Peter visited and also sent a greetings message to VK3GEE on 15m.

M CALLSIGNS

From 1 April 1996, a new type of amateur radio callsign was issued in the UK using the M prefix instead of G. Starting with M0 and M1 for the Class A and B licences respectively. At a ceremony held on 30 March, the first callsigns were presented. M0AAA went to Reading and District ARC and M1AAA went to Ian Oliver.

The RSGB Patron visited GB0OSH at Great Ormond Street Hospital.

MORE ECLIPSE EXPERIMENTS

The total solar eclipse on 11 August 1999 provided extraordinary propagation, particularly on the lower frequency amateur bands. Ray G4FP, Alan GI3WWM and John G0LLT recorded interesting results shortly before the moon's shadow reached the UK. Having informed a number of Canadian and US stations of their intent, they put out "CQ DX" calls on 3799kHz SSB from about 1000UTC on the 11th. At 1015UTC, both John and Ray heard a weak signal responding but due to the UK and European interference no two-way contact resulted. Later it was confirmed that Rob VY2ROB, Barry VO1BAR/P and Rolf KE1Y had all heard the signal from G4FP. Transatlantic propagation on 80m after 1000UTC in summer is extremely rare. On 160m, Mike G3SED reported enhanced propagation for about 20 minutes either side of the period of totality with much enhanced signals for the four or five minutes around totality. Mike heard, or worked, ON4UBA, IV3PRK, GM3POI in Orkney and F/G3TKN in central France. LF enthusiasts monitored time signal station HBG in Switzerland on 75kHz. Stations in the UK, Netherlands and Germany all noted a sudden appearance of sky waves that interfered with the normally audible ground wave signal from HBG.

DATAMODES

Peter Martinez G3PLX has played a significant role in the development of what are known to most nowadays as amateur datamodes. Having started on this path back in the 1960s by making extensive use of RTTY, he was instrumental in bringing AMTOR to amateur radio by the end of the 1970s. This development dramatically increased the reliability of HF radio links and paved the way for a number of advanced new modes to support data transfer and email systems. However, by the 1990s he realised that, despite all the advances, no system had been developed to improve live, keyboard-to-keyboard, contacts. He therefore set about designing a completely new system to satisfy this need at the end of 1999. By taking a fresh look at the requirements he was able to use the technical developments of the time to create a near perfect system. By using phase shift keying and a bandwidth of just over 30Hz, the new system provided a remarkably robust link with a transmission speed faster than most amateur operators could type. The narrow bandwidth meant that the entire activity for a busy band such as 20m could be contained in a single voice channel. The mode, of course, was PSK-31. Today, PSK-31 is arguably the most popular digital mode in the world and the original design was so good that no further changes have been necessary.

Following awards from both the RSGB and ARRL for his work on digital communications, in 2012, he was awarded the YASME Excellence Award for his distinct contribution to digital communications technology. Peter was recognised for his invention of PSK-31, a widely used digital mode that enables many amateurs to successfully communicate on HF with very modest stations.

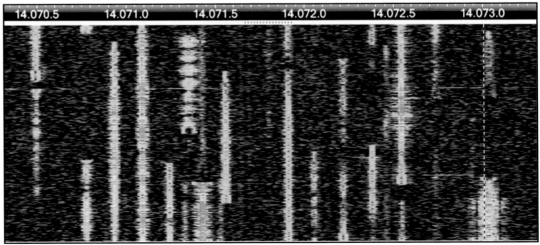

Waterfall display of PSK-31 activity on 20m.

Centenary

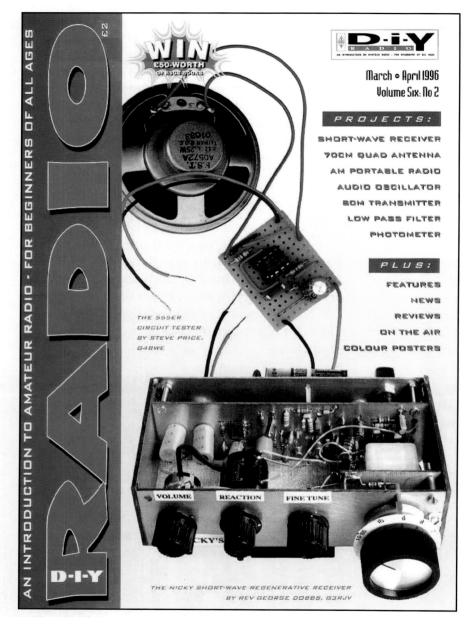 

Two other magazines that the RSGB produced in the 1990s.

6

New Millennium

WELCOME TO THE 21ST CENTURY

The first *RadCom* editorial in the new millennium said, "Welcome to the 21st century. With the Millennium celebrations behind us we can all look forward to the next one hundred years of technical development. I wonder if the founders of the Society ever gazed into the crystal ball and wondered where radio communications would be in the year 2000? From humble beginnings at the turn of the last century, radio communications has developed further and faster than they probably would have envisaged. What of the next one hundred years? Certainly amateur radio played a key role in moving communications technology on in the last century and I am sure that we have a vital role to play in the future of communications in this, the 21st century. Are we up for the challenge? I believe we are, certainly the Society is gearing up to play a continuing role in representing amateur radio."

To see in the new Millennium, Cray Valley Radio Society ran M2000A starting on 31 December 1999. One of the five stations concentrated on making prearranged contacts in over 30 DXCC entities as they entered the new millennium. Lord Rix G2DQU made the first QSOs including one to ZL6A at 1100UTC, their midnight.

HEDY LAMAAR

Hedy Lamaar, the Hollywood leading lady of the 1930s and 1940s, who held the original patent for the technique that we now know as spread spectrum, died in 2000. A better example of lateral thinking is hard to find. She was a friend to the composer George Antheil and used to spend time playing piano duets with him. They played a game in which one of them would suddenly change key and the other would have to follow as quickly as possible. In the period when both were in different keys, the music would be unintelligible, making sense only when the keys were the same. Hedy had the breadth of knowledge (and the undisputed intelligence) to realise that changing key on the piano was a form of frequency-hopping, and when the players were in different keys, nothing made sense. She went on to determine that a form of communications could be devised whereby the transmitter and the receiver kept 'changing key' – changing frequency synchronously – so that they always kept in step with each other. Such a system would be very secure, both from the aspects of being intercepted and being jammed. She was granted a patent for the technique in 1942 and was also awarded the Electronic Frontier Foundation Prize for the secure, high-bandwidth communications.

REGIONAL STRUCTURE

In 2000, the RSGB proposed a new geographic representational structure. At the time the UK was separated into seven Zones and each of these Zones was represented on the RSGB Council. It was felt that these Zones were too large to keep in touch with the grassroots. Under the new proposals, the Zones would be increased to 12 and be renamed Regions. Each Regional manager would be assisted by deputies. They would, as a team, be responsible for all Society activity within their respective Region. This included club and school liaison, attendance at local rallies and events and Membership recruitment as well as PR at a local level.

The changes started on 20 January 2001 following acceptance at the 2000 AGM.

HANDHELD INVENTOR SK

Al Gross W8PAL died just before Christmas 2000, aged 82. He pioneered the development of radio transmissions above 100MHz. He obtained his amateur licence in 1934 at the age of 16 and his early interest in amateur radio set his career choice while he was still a teenager. In 1938, while still at college, he invented the first handheld radio transceiver, which he called a 'walkie-talkie'. Gross was recruited by the OSS, the forerunner to the CIA, where he developed two-way air-to-ground communications used behind enemy lines during WWII.

GB4FUN

The RSGB took delivery of a vehicle to be used as an amateur radio demonstration station in early 2001. The vehicle had been donated to the Society by the Radiocommunications Agency. The handing over of the keys took place at the Baldock Monitoring Station when the vehicle was formerly based. The vehicle was then

The first GB4FUN in use.

6. New Millennium

converted into a demonstration station after which it began a programme of school visits throughout the UK aimed at promoting the use of amateur radio in schools. Called GB4FUN, the vehicle made its first operational public appearance at the Bedfordshire Steam Fayre in September 2001. It had a state of the art amateur station and a 25ft telescopic mast. GB4FUN also appeared at the Leicester Amateur Radio Show, the W&S Open Day in Glenrothes and the HF Convention. After these visits it was fitted with a packet station, satellite communications, an IT suite and video equipment in preparation for its full operation programme the following year.

In March 2002, GB4FUN spent the day at Rydal Penrhos School in Colwyn Bay. Alongside the North Wales Radio Rally Club, teacher Anton MW1EYT set up an HF station in the school lab. They demonstrated the use of computers

The first GB4FUN vehicle, interior.

The second GB4FUN, interior.

The second GB4FUN.

in amateur radio to the pupils and the club members gave short talks and answered questions. Later in the year it was reported that six boys from the school had taken their Foundation Licence as a direct result of the visit to the school by GB4FUN. These were just the first pupils, scouts and guides who went on to do the Foundation Licence following a GB4FUN visit.

Fund raising started in 2006 for a second GB4FUN vehicle and in 2008 the RSGB took delivery of the new GB4FUN trailer built as a mobile classroom fitted out with a wide range of amateur equipment. It was large enough to deliver talks and allow pupil interaction, but also compact enough to park up at all the schools with minimal fuss and disruption. Its facilities aimed to provide a wonderful resource to bring amateur radio into the lives of many more young people. The new GB4FUN trailer had its first outing at the RSGB HF Convention in October 2008, followed by a public launch at the Leicester Amateur Radio Show. Unlike previous years, where its predecessor was outside in the flea market, this time it was a centrepiece of the RSGB presence. Much of the equipment inside the vehicle was donated or loaned by the major equipment suppliers, groups and individuals. But the trailer itself could not have been purchased without the support of the Radio Communications Foundation.

CONTESTS CANCELLED

Whilst the Foot & Mouth Disease Declarations (Controlled Area) Order 2001 was in force, the RSGB advised all UK radio amateurs to avoid any activity that would bring them into contact with farmland or animals susceptible to Foot and Mouth Disease (FMD), such as pigs, cattle, sleep and goats. This mainly applied to portable contest operating, rallies, ARDF events and maintenance visits to remote repeater, beacon and packet sites. In the light of those Foot & Mouth Disease restrictions, the RSGB VHF Contest Committee announced that the Portable sections of all its contests were suspended. The RSGB HF NFD was cancelled as were the Portable sections of the Low Power Contest in July. This temporary rule remained in force until the outbreak had been controlled. Portable contests resumed in September.

NANO METRE RADIO WAVES

Three amateurs teamed up to make the first laser QSO between England and Wales in March 2001. Steve GW4ALG, Paul G0ONA and David G0MRF used modulated laser pointers at the heart of two CW transceivers operating at 670nm. On a day frustrated by heavy rain and access to suitable sites being prevented by Foot & Mouth restrictions, the first QSO was made between GW0ONA/P and G0MRF/P across the River Wye – a modest distance of 100m. Having made a QSO, the group moved on to the River Severn where the path length was a more challenging 5.1km. After 20 minutes, a two-way QSO was completed between GW0MRF/P who received a 589 report and G0ONA/P who was 569 on the Welsh side of the border.

NEW LICENCE STRUCTURE

In October 2001 the start of a new 3-tier licence system was announced by the Radiocommunications Agency – the Foundation Licence being amongst a series of changes to take effect 1 January 2002. This was a prelude to a revised integrated structure of qualification and exams for amateur radio planned for final implementation in 2004. The Novice licence was re-named the Intermediate licence and the Class A, A/B and B licences were re-named the Full licence. Holders of the new Foundation Licences were issued callsigns with the prefix M3. Figures from the RA show that in the first three months of the year, 1,407 M3 callsigns had been issued. Of those, 313 were candidates of the new course and exam, the remaining were Class B licensees who, on completion of a Morse test, obtained the new licence with its HF privileges. A new Intermediate Licence syllabus was

Centenary

The GB50 special event station for the Queen's Golden Jubilee.

introduced in early 2003 and the new Full Licence syllabus in 2004. Now, entry was exclusively via the Foundation Licence.

QUEEN'S GOLDEN JUBILEE

To celebrate the Queen's Golden Jubilee, the Radiocommunications Agency issued GB50 for a special event station held at Windsor Castle. The station was run by the Cray Valley Radio Society, Burnham Beeches Radio Club and the RSGB. The Patron of the Society, His Royal Highness, Prince Philip, Duke of Edinburgh, KG, visited the station on Monday 3 June 2002. He spent some time viewing the display of historic radio equipment, including a Marconi transmitter dating from 1901, before being shown the current state-of-the-art equipment used at GB50. He listened

The Patron visiting the GB50 special event station.

to an SSB contact with 9H1EL in Malta and was shown a CW contact too. A special Golden Jubilee greetings message from His Royal Highness was transmitted, "As Patron of the Radio Society of Great Britain, I am delighted that it has been able to set up the GB50 special event station on the North Terrace of Windsor Castle overlooking the Thames and the town of Windsor. It is in a very appropriate position to receive messages of good wishes from amateur radio enthusiasts to the Queen in her Jubilee Year. I know the Queen very much appreciates this special contact with people throughout the Commonwealth, and the rest of the world, and she has asked me to send you all her warm thanks for your support and affection as this time. I hope that all your contacts with GB50 over the next ten days will be five and nine. 73. Philip". Around 5,000 members of the public visited the exhibition and nearly 500 of those were able to send greetings messages over the air.

5MHz EXPERIMENTS

In 2002, permission was granted by the MoD and the RA for the allocation of five spot frequencies between 5.260 and 5.405MHz. The purpose of the experiment was to carry out propagation and antenna investigations aimed at improving the understanding of Near Zenithal Radiation or NVIS (Near Vertical Incidence Skywave) communication via the ionosphere. Those interested had to apply for a Notice of Variation to their amateur licence. In 2005, GB3ORK commenced operation from Orkney in support of the 5MHz experiment. Reception reports of the beacon were sent to the 5MHz Working Group.

END OF MORSE REQUIREMENT

The Radiocommunications Agency published a Gazette notice in July 2003 announcing the end of the Morse requirement for access to the HF bands in the UK. Full and Intermediate Class B Licensees automatically had Class A privileges and were allowed to operate on the HF bands with their existing callsign.

RADIO COMMUNICATIONS FOUNDATION

The Radio Communications Foundation (RCF), the RSGB's charitable arm, was created in 2004. The aims of the Foundation were "to advance the education of the public in the science and practice of radio communication and electrical engineering and to promote her benefits to the public resulting from such education and training". In 2004, the RCF joined forces with the Arkwright Trust to provide bursaries to 6th Form students who elect to take maths and a science-based subject at 'A' level and who have an interest in radio communications. Then, in 2005, the RSGB and the RCF took on the responsibility for licensing examinations in the UK. The RSGB took over the day-to-day administration and the RCF accredit the exams.

Centenary

PARTY IN THE PARK

The RSGB celebrated the 90th anniversary by encouraging Members to get involved in the "Great Birthday Party". The whole of the UK really got involved with the festivities with Party in the Park. Special event stations based on GB90RSGB were put on all over the country showcasing amateur radio to the general public over the weekend 26 and 27 July. Clubs organised everything from HF and VHF stations to ATV demonstrations and car boot sales. The venues were varied too – museums, parks, community centre and even railway stations and the BT Transmitting Station at Rugby.

6. New Millennium

ARDF NATIONAL CHAMPIONSHIPS

The first ever Amateur Radio Direction Finding (ARDF) National Championships in the UK took place at Ash Ranges near Aldershot in November 2006. Direction finding is the amateur radio equivalent of orienteering; contestants use direction finding radios to find the location of a series of transmitters dotted around a piece of countryside. The transmitters are all on the same frequency and transmit with a keyed ID in a strict sequence over a five minute period. This then repeats over and over again for the duration of the competition. The winner is the person or team who finds all the transmitters in the shortest space of time. The first British Championships followed the format of ARDF events in the Netherlands, with the number of transmitters that participants had to find being determined by their age. Younger contestants were set the task of hunting five transmitters while older participants only had to find three or four.

First place was taken by David Williams RS190108, who hunted his allocated four transmitters down in 1 hour 13 minutes and 45 seconds. Michael Dunbar RS195082, was runner-up and one of the German visitors, Dirk Smit DH1YHU, came in third.

Two competitors line up at the start. Photo David Williams.

The group of German amateurs who travelled over for the event. Photo G3ORY

David Williams receiving the 144MHz salver from Dave Wilson M0OBW. Photo G3ORY

Centenary

Thilo Kootz DL9KCE, Sotirios Vanikiotis SV1HER, Jacques Mezan de Malartic F2MM, Gaston Bertels ON4WF, Don Field G3XTT, John Pink G8MM, Dennis Hartig DL7RBI.

EUROPEAN PARLIAMENT EXHIBITIONS

The RSGB was represented at a Eurocom/IARU exhibition of amateur radio at the EU Parliament in Brussels in 2007. The exhibition was a showcase to MEPs, European Commission staff and others, highlighting the benefits of the hobby to the wider community. It focused on technical training, careers and emergency communications. The German society DARC also had a demonstration of the interference caused by various household devices and power line communications. The RSGB was represented by the President, General Manager and a small team of volunteers who, along with colleagues from DARC, REF and UBA, the German, French and Belgian national radio societies, stayed on to man the stand for the remainder of the week. Overall, the exhibition was a useful opportunity to influence those who draft Europe-wide legislation including policies on matters such as EMC, particularly in respect of the open market.

A second exhibition in 2010 looked at the amateur radio service and its structures and benefits for society, including emergency communications during times of national disaster. Using the motto European Amateur Radio Benefiting Society, the event was sponsored by the IARU Region 1 EUROCOM Working Group and European Parliament Member (MEP) Birgit Sippel, who supports the goals and the importance of amateur radio. The RSGB has a prominent role in the work of EUROCOM. "Even though the amateur radio service has been around since 1908 and many countries even have special laws to regulate it, the service is often unknown in public," said EUROCOM Working Group Chairman Thilo Kootz DL9KCE. "In the European Union alone, about 350,000 people of all ages are fascinated by this hobby. A combination of communication, technology and sports bonds them together and makes amateur radio unique." During the exhibition, 10 students from a school in Brussels contacted the International Space Station.

International Space Station veterans
Roman Romanenko of Russia, Canadian Robert Thirsk and Belgian Frank De Winne

The space communications and amateur radio's role in disaster communications exhibition.

Centenary

FRIEDRICHSHAFEN ADDRESS

RSGB President Angus Annan MM1CCR was given the rare honour of making the opening address at the annual Friedrichshafen radio show in Germany in June 2007. His speech, delivered to an invited audience and relayed to a giant screen, covered the recent successes and tribulations of the amateur radio scene. After the opening ceremony he attended a number of high level meetings.

6. New Millennium

AGM BROADCAST LIVE

For the first time ever, the RSGB 2009 AGM was broadcast live on the internet thanks to the British Amateur Television Club (BATC) from the Novotel Newcastle Hotel on Saturday 18 April. Using their established live web streaming technology, volunteers from the BATC captured events and played them live from their streaming site. There was also a live web chat facility on the live stream page. After the formal part of the meeting, the trophies and awards were presented.

The first AGM to be broadcast live over the internet.

TALKING TO THE ISS

Budbrooke School not only had the opportunity to talk to the International Space Station in October 2008, they actually got to meet the astronaut they spoke to, Richard Garriott, W5KWQ.

When the ISS was well and truly over the Atlantic, the GB-4FUN team started calling a few minutes before the ISS was over the visible horizon in the hope that they would acquire the ISS early. On about the eighth or ninth call, a voice responded – it was Richard and the contact was on. Some eight months later, Richard Garriott travelled to Warwickshire to meet the children, teachers and parents of Budbrooke School. He first met with the youngsters who had asked the questions on the day of the actual contact, where they had the chance to put other questions to him about his time in space. He then went on to speak to all the children at their school assembly.

6. New Millennium

The original contact with Richard Garriott W5KWQ onboard the ISS.

Richard Garriott visits the school to speak to the children.

NATIONAL RADIO CENTRE

Work commenced in early January 2010 on the RSGB's National Radio Centre (NRC), based at Bletchley Park. The Centre is a world-class showcase for radio communications technology as a force powering the 21st century economy and to present amateur radio as an exciting, stimulating, educational, multi-faceted hobby, which provides a sound technical grounding in radio communication for those within its ranks. Amateur radio is fun, fast moving and within reach of anyone who can undertake a short period of training. Whilst the NRC is not a museum there is a legacy area where outstanding examples of amateur radio equipment through the 20th century are exhibited. There is also a library reference area and archive, access to which is on a by appointment basis to allow those who wish to research the extensive archives that the RSGB has built up over

Breaking ground for the RSGB's new National Radio Centre

nearly 100 years to do so. Initially, the NRC was only open at weekends during early 2012 whilst any early 'teething issues' were resolved. By the time it was officially opened by Ed Vaizey MP, Minister for Culture, Communications and Creative Industries in July 2012, opening hours had been extended.

On unveiling the plaque in the entrance, Ed Vaizey, MP said, "The National Radio Centre will do an enormous amount to spread the knowledge about the history of radio and its continuing importance in the 21st century. The RSGB is, technically, populated by amateurs but it is actually full up to the brim with technical professionals who have done an amazing job here. What really appeals to me about the work that the RSGB does is the importance of introducing young people to technology. It gives them great purpose, with the ability to build something and then have the possibility to talk to someone on the other side of the globe, which can fill them with excitement. The NRC is a great reminder of Britain's place in the history of radio and the incredibly important role that Britain has played in developing radio and we still pioneer not just in the field of amateur radio but also in the field of digital radio."

Some of the interactive displays.

Ed Vaizey, MP opens the National Radio Centre.

GB3RS encourages visitors to send greetings messages around the world.

SPECTRUM DEFENCE FUND

The Spectrum Defence Fund was launched in January 2010 by RSGB President Dave Wilson M0OBW. This fund was used, in the first instance, to consider a challenge to Ofcom at law with regards to Ofcom's interpretation of the various acts and directives that cover PLA/PLT and the threat that they pose. The RSGB and other EMC specialist groups had been unhappy with progress and regard the Draft developed as providing wholly inadequate protection to the HF spectrum. The Draft Standard was put out to vote and RSGB and IARU ensured that all European Member Societies were in a position to provide informed input to their National Standardisation Organisations prior to voting. In the UK a substantial number of members of the BSI committee (including RSGB) were opposed to the draft and the Committee abstained from supporting the draft standard.

The Society believes that very significant threats to the radio spectrum remain. The encroachment of digital technologies into domestic environments and the general reduction of dependency on HF communications, taken together with increasing demand for spectrum for other services, are constitute an increasing pressure on the interests of the amateur services.

QUEEN'S DIAMOND JUBILEE & OLYMPIC GAMES

The summer of 2012 was a busy one for amateurs in the UK. The RSGB and Ofcom reached agreement on the optional use of special callsigns for two significant events. For the Queen's Diamond Jubilee, all UK stations had the facility, if they wished, to use the letter 'Q' in the place of the Regional identifier in the callsign. For the period of the Olympic Games and Paralympic Games, a similar facility

2O12L on the air.

existed for a seven week period, using the letter 'O'. Special Event Callsigns in the series GB2012aaa were issued for 'flagship' stations using the prefix 2O12a where 'O' is the letter O, and a is a single letter relating to the location of the station.

Just like their athletic counterparts, the organisers of the two stations worked hard for a long time to make sure that the stations were a credit to the hobby. Barry Amateur Radio Society and Cray Valley Radio Society put the two flagship events on the air supported by the RSGB. The Olympic Games puts the UK in the spotlight with its vast world-wide TV and radio audiences and the thousands of visitors who came to share in the experience. The Olympic stations showcased amateur radio, not just to the many thousands in the log, but to the public who visited.

During the Olympic Games, several radio amateurs joined Ofcom to use their skills in radio engineering and interference to support the needs of the Games as Ofcom didn't have enough people from their own resources for the task.

Ofcom said, "Your support for this initiative introduced us to people who were very well motivated and with a high level of technical skill and expertise. We could not have provided such a good service to our customers without their help. We had the benefit of learning from the experience and knowledge that they brought."

GOLD MEDAL WINNER

God Save the Queen was heard for the first time at the 2012 World Amateur Radio Direction Finding Championships following the victory of Bob Titterington G3ORY in the 80m classic race. This was the very first podium finish for the RSGB since the sport of IARU format Amateur Radio Direction Finding started in the UK back in 2002.

QSL card with both NoVs for the Queen's Diamond Jubilee and the Olympic and Paralympic Games.

Bob Titterington G3ORY wins the UK's first ARDF gold medal.

Centenary

RSGB CENTENARY YEAR

Centenary Year, 2013, started with two new amateur bands allocated to UK radio amateurs. The new 600m band, 472 – 479kHz, was available to UK amateurs holding Full licences with a Notice of Variation (NoV). The second new band was 5MHz. Although access to 5MHz in the UK had been available on spot frequencies, this changed to 11 individual band segments of varying width allocated to Amateur (Full) Licence holders under an NoV.

A number of events to mark the RSGB's Centenary were in place for 2013. The special event callsign G100RSGB travelled the country spending 28 days in each region. Local radio clubs and special interest groups worked thousands of stations, encouraging activity during the Centenary year. Other activities including celebrations on 5 July at the National Radio Centre in Bletchley Park, a Centenary Construction Competition and world-wide award schemes.

Gx100RSGB toured the country through all 13 Regions.

Graham Coomber G0NBI, General Manager, said, "Whilst Centenary celebrations are a time to reflect on our past achievements, there can be no question of "resting on our laurels". Amateur radio today is as relevant and vibrant as it has ever been, but we face new challenges such as encouraging new entrants, combating new sources of interference and ensuring that our voice is heard and understood, as pressure on the radio spectrum increases. The RSGB takes very seriously its responsibility to work with partners around the world in providing leadership to safeguard and develop amateur radio over the next 100 years."

RSGB Annual General Meeting 2013, held at the IET, Savoy Place, London.

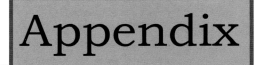

100 Years of RSGB Presidents

1913-20	Alan A. Campbell Swinton, 2HK	1959	Reginald Smith-Rose, CBE	1983	Donald Baptiste, CBE
1921	Major John Erskine-Murray, MUX	1960	William Metcalfe, G3DQ	1984	Robert Barrett, GW8HEZ
1922	Admiral of the Fleet Sir Henry Jackson, KCVO	1961	Major-General Eric Cole, CBE, G2EC	1985	Joan Heathershaw, G4CHH
1923-24	Professor W.H. Eccles, EWX, 2BA	1962	Edward Ingram, GM6IZ	1986	Willie McClintock, G3VPK
1925	Sir Oliver Lodge, FRS, SCc, LLD,	1963	Norman Caws, G3BVG	1987	Joan Heathershaw, G4CHH
1923-27	Brigadier General Sir Capel Holden, KCB	1964	Geoffrey Stone, G3FZL	1988	Sir Richard Davies, KCVO, CBE, G2XM
1928	Captain Ian Fraser, G5SU	1965	Eric Yeomanson, G3IIR	1989	Julian Gannaway, G3YGF
1929-30	Gerald Marcuse, G2NM	1966	Roy Stevens, G2BVN	1990	Frank Hall, GM8BZX
1931-33	Henry Bevan Swift, G2TI	1967	Alexander Patterson, GI3KYP	1991	Ewart Case, GW4HWR
1934-36	Arthur Watts, G6UN	1968	John Graham, G3TR	1992	J. Terence Barnes, GI3USS
1937	Ernest Dawson Ostermeyer, G5AR	1969	John Swinnerton, G2YS	1993	Peter Chadwick, G3RZP
1938-40	Arthur Watts, G6UN	1970	Dr John Saxton	1994	Ian Suart, GM4AUP
1941-43	Alfred Gay, G6NF	1971	Frederick Ward, G2CVV	1995	Clive Trotman, GW4YKL
1944-46	Ernest Gardiner, G6GR	1972	Robert Hughes, G3GVV	1996	Peter Sheppard, G4EJP
1947	Stanley Lewer, G6LJ	1973	Dr John Saxton	1997-98	Ian Kyle, GI8AYZ
1948-49	Victor Desmond, G5VM	1974	George Jessop, G6JP	1999	Hilary Claytonsmith, G4JKS
1950-51	William Scarr, G2WS	1975	Cyril Parsons, GW8NP	2000-01	Don Beattie, G3BJ
1952	(Dud) F J H Charman, BEM, G6CJ	1976	Ernest Allaway, G3FKM	2002-03	Bob Whelan, G3PJT
1953	Leslie Cooper, G5LC	1977	Lord Wallace of Coslany	2004-05	Jeff Smith, MI0AEX
1954	Arthur Milne, G2MI	1978	Dain Evans, G3RPE	2006-07	Angus Annan, MM1CCR
1955	Herbert Bartlett, G5QA	1979	John Bazley, G3HCT	2008-09	Colin Thomas, G3PSM
1956	Reginald Hammans, G2IG	1980	Peter Balestrini, G3BPT	2010-13	Dave Wilson, M0OBW
1957	Douglas Findlay, G3BZG	1981	Basil O'Brien, G2AMV	2013-14	Bob Whelan, G3PJT
1958	Leonard Newham, G6NZ	1982	Ernest Allaway, G3FKM		

INDEX

Alford, Ken (TXK) G2DX	30	
Allen, Bert G8IG	67	
Amateur Television	49, 50, 51	
ARDF	129, 139	
ARRL	62, 64	
Charman, Dud G6CJ	68, 69	
Clarricoats, John G6CL	52, 63	
Cooper, Leslie G5LC	80	
Corry, Nell G2YL	15, 16, 17, 20	
	61, 63, 65	
Datamodes	89, 115	
Desmond, Victor G5VM	60	
Dunn, Barbara G6YL	15, 22, 42, 63	
Eccles, Professor W H	29	
EMC	102	
Emergency Comms (RAYNET)	54, 77, 78, 79,	
	80, 110	
European Parliament	130, 131	
Exhibitions	75, 81	
Falklands War	104	
Fogarty, L Francis (FFX)	1, 3, 4	
Gay, A D G6NF	18	
GB0OSH	114	
GB1RS	73	
GB2LO	96, 97	
GB2RS	83	
GB2SM	82	
GB4FUN	122, 123, 124	
Golden Jubilee	94	
Goyder, Cecil G2SZ	10, 11	
Hall, Constance G8LY	17, 20, 94	
Hawker, Pat G3VA	88	
Heathershaw, Joan G4CHH	107	
Heightman, Denis G6DH	17, 72	
Holmes, Robert G6RH	73	
Hope-Jones, Frank	1, 3, 4	
International Amateur Radio Union (IARU)	34, 35, 75, 89, 90, 103	
International Geophysical Year	86	
Islands on the Air (IOTA)	95	
Jessop, George G6JP	47	
Karlson, John (SMUA) SM6UA	26	
King Hussein of Jordan JY1	101	
Klein, Rene (RKX) G8NK	1, 2, 3, 4	
Lewer, Stanley G6LJ	74	
Lodge, Sir Oliver	28	
London Wireless Club	1, 4	
Marcuse, Gerald G2NM	13, 14, 23, 24, 34, 36, 37, 63, 69	
McMichael, Leslie (MXA) G2MI, G2FG	1, 2, 3, 4, 45	
Morgan, A P	1, 3	
National Field Day	43, 70, 71, 92, 93, 105, 125	
National Radio Centre	136, 137,	
Newnham, Len G6NZ	57, 99	
Novice Licence	113	
Olympic Games	138, 139	
Ostermeyer, E Dawson G5AR	19	
Patrons	31, 32, 108, 109, 112, 114, 114, 126	
Pope, H W (PZX)	6, 22	
Radio Communications Foundation (RCF)	127	
Radio Society of Great Britain	7	
Rallies	83, 84	
Repeaters & Beacons	99	
Royle, Ralph G2WJ	20, 21, 36, 38, 49, 50, 51	
RSGB Centenary	140	
Satellites/Space	87, 90, 91, 101, 102, 106, 107, 110, 111, 134, 135	
Scarr, William G2WS	76	
Science Museum Lectures	98	
Simmonds, Ernest G2OD	5, 8, 9, 11, 12, 13, 14, 42	
Southern Cross	15, 40, 41	
T & R Bulletin - RadCom	38, 39, 40, 56	
T & R Section	33	
Technical Topics	87	
Transatlantic tests	7, 23, 30, 33, 72, 90, 103	
Voluntary Interceptors	61	
Walters, Douglas G5CV	46, 47,	
WARC	100	
Watts, Arthur G6UN	53	
World War 1	5	
World War II	55	
Young Amateur of the Year (YAOTY)	108	